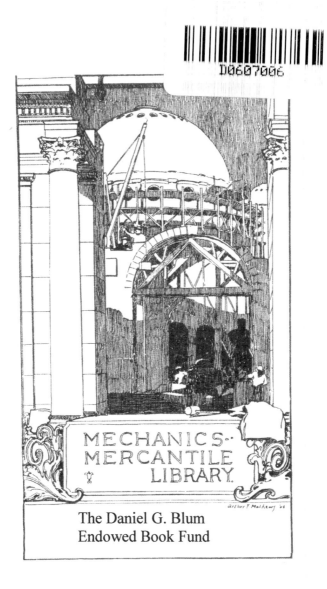

Advance Praise for *Satellites in the High Country*

"Jason Mark revisits 'the wild' in our landscapes and in our minds. At a time when the wild—as a place and an idea—is being increasingly hemmed in, he offers fresh insights, unsettled questions, and renewed appreciation."

> —**Curt Meine**, Aldo Leopold Foundation
> and Center for Humans and Nature

"In *Satellites in the High Country*, gripping accounts of outdoor journeys are linked with provocative thinking about the meaning of wildness in an increasingly human-controlled world. Jason Mark ably continues the writing style and themes of legends such as John Muir and Edward Abbey."

> —**Roderick Frazier Nash**, Professor Emeritus of History and
> Environmental Studies, University of California Santa Barbara
> and author of *Wilderness and the American Mind*

"*Satellites in the High Country* is an act of ground truthing on the nature of wildness at this moment in time. Author Jason Mark circumnavigates the American West with the eyes of an open-hearted sleuth, looking for what wild remains. Wildness, he discovers, is not only all around us, but inside us as well, having little to do with what is pristine or untouched and everything to do with nature's intricate system of adaptation and response, function and beauty, and our innate capacity for awe. This book is a conversation with sanity."

> —**Terry Tempest Williams**, author of *When Women Were Birds*

"Jason Mark is a great person to share an adventure with, whether out on the Arctic tundra or on the page. *Satellites in the High Country* is an engrossing exploration of the ever-evolving definition of what is 'wild' in America—which often reveals as much about us as it does about wilderness in the twenty-first century."

> —**Michael Brune**, Executive Director, Sierra Club

"*Satellites in the High Country* is a brave and vigorous exploration of wilderness—its meaning, its necessity, its thunderous, rock-strewn reality. Jason Mark guides the reader across mountain passes and Arctic tussocks on a journey that is at once physical, philosophical, and political. His feet may be bruised, but his voice is strong, honest, and compelling. Read this book for an insightful and much-needed update on the centrality of wilderness in the contemporary American mind."

> —**Kathleen Dean Moore**, author of *Great Tide Rising*

About Island Press

Since 1984, the nonprofit organization Island Press has been stimulating, shaping, and communicating ideas that are essential for solving environmental problems worldwide. With more than 800 titles in print and some 40 new releases each year, we are the nation's leading publisher on environmental issues. We identify innovative thinkers and emerging trends in the environmental field. We work with world-renowned experts and authors to develop cross-disciplinary solutions to environmental challenges.

Island Press designs and executes educational campaigns in conjunction with our authors to communicate their critical messages in print, in person, and online using the latest technologies, innovative programs, and the media. Our goal is to reach targeted audiences—scientists, policymakers, environmental advocates, urban planners, the media, and concerned citizens—with information that can be used to create the framework for long-term ecological health and human well-being.

Island Press gratefully acknowledges major support of our work by The Agua Fund, The Andrew W. Mellon Foundation, Betsy & Jesse Fink Foundation, The Bobolink Foundation, The Curtis and Edith Munson Foundation, Forrest C. and Frances H. Lattner Foundation, G.O. Forward Fund of the Saint Paul Foundation, Gordon and Betty Moore Foundation, The JPB Foundation, The Kresge Foundation, The Margaret A. Cargill Foundation, New Mexico Water Initiative, a project of Hanuman Foundation, The Overbrook Foundation, The S.D. Bechtel, Jr. Foundation, The Summit Charitable Foundation, Inc., V. Kann Rasmussen Foundation, The Wallace Alexander Gerbode Foundation, and other generous supporters.

The opinions expressed in this book are those of the author(s) and do not necessarily reflect the views of our supporters.

Satellites in the High Country

Satellites in the High Country
Searching for the Wild
in the Age of Man

Jason Mark

◗ ISLANDPRESS Washington | Covelo | London

Island Press is a trademark of The Center for Resource Economics.

Library of Congress Control Number: 2015941186

Printed on recycled, acid-free paper ♲

Manufactured in the United States of America

10 9 8 7 6 5 4 3 2 1

All map illustrations by Kristin Link (www.kristinillustration.com).

Lyrics from "THE GLOAMING" by THOMAS EDWARD YORKE, PHILIP
JAMES SELWAY, EDWARD JOHN O'BRIEN, JONATHAN RICHARD, GUY
GREENWOOD and COLIN CHARLES GREENWOOD appear with permission
from Alfred Music Publishing. © 2002 WARNER/CHAPPELL MUSIC LTD (PRS).
All Rights Reserved.

Keywords: Island Press, Anthropocene, The Human Age, The Age of Man, wilderness,
wild, wildness, Aravaipa Canyon, Yosemite, Olympic Mountains, Hoh Rainforest,
Olympic National Park, Aichilik River, Arctic National Wildlife Refuge, Gwich'in,
Badlands, Black Hills, Lakota, Gila Wilderness, pristine, Mexican Gray Wolf, rewilding,
Lynx Vilden, Stone Age Living Project, Google Treks, Columbian Exchange,
wildling, naturalness, geoengineering, climate change, untrammeled, Outward
Bound, Pine Ridge Reservation, powwow, Aldo Leopold, John Muir, the land ethic,
The Wilderness Act, William O. Douglas, Henry David Thoreau, Charles Darwin,
evolution, conservation, conservationists, environmentalists, ecology of fear, Yosemite
National Park, Shelton Johnson, frontier, Theodore Roosevelt, America's Best Idea,
Drakes Bay Oyster Farm, Point Reyes National Seashore, William Cronon.

For those who cannot live without wild things

What would the world be, once bereft
Of wet and wildness? Let them be left,
O let them be left, wildness and wet;
Long live the weeds and the wilderness yet.

— Gerald Manley Hopkins, "Inversnaid"

Contents

Aravaipa Canyon

Prologue

Into the Wild

M Y FOOT WAS KILLING ME. As long as I was able to step flat and keep my heel and toes level, the pain wasn't too bad. But walking evenly was impossible in that mostly trackless wilderness. I kept losing the trail, picking it up again, blazing my own. I stumbled over river rocks, mud patches, deadfalls, thickets of branches—the natural mess made by the flash floods that sometimes tear through Aravaipa Canyon, Arizona.

Whenever my toes bent, the inch-long piece of wood lodged deep between the skin and bone of my left foot stabbed into me. That really hurt. The swelling was much worse. Overnight my foot had bloated into something resembling an overstuffed sausage, and as I tried to make my way out of the canyon it throbbed incessantly.

I don't want sound melodramatic about the whole thing. Yes, I was in a rough spot—miles away from assistance, and hurt. But it wasn't like that guy who got his arm trapped under a boulder and had to cut off his hand. I had told several people where I was going, the exact trailhead, and my expected departure from the backcountry. It wasn't as if I were going to die.

Still, I was nervous. I had been in the canyon a few days and had seen only one pair of hikers, who had been headed back out. There was no one around to assist me, no one to hear a cry for help even if I made one. I looked up at the salmon-pink cliffs towering hundreds of feet above and knew, with a twist of fear, that my rescue would have to be my own. Suddenly I felt very vulnerable. What was supposed to have been a fun adventure had turned dangerous. All of my energies had been distilled to a single, primal motive: getting out of there in one piece.

I leaned on the cottonwood crutch I had made the day before and took in the stillness of the canyon, its unremitting silence. For about the twentieth time that morning I pulled the canyon map from my back pocket to see where I was. I looked upstream. I looked downstream. I tried to measure how far I had gone, how far I still had to go. At least six miles, probably more, and every other step was guaranteed to hurt like hell.

There's no such thing as bad weather, only poor gear decisions. A rainstorm is a delight if you've got a good slicker. Gear, of course, includes footwear. In hindsight, it's obvious that wearing a pair of sandals on a rugged and often nonexistent trail wasn't the best idea. At the time, though, it seemed to make sense.

Aravaipa Canyon is extremely narrow—at many points, probably no more than a quarter of a mile from rim to rim—which means that to explore the canyon you often hike right through the streambed. Traverse the entire twelve-mile length of the canyon and you'll cross the creek at least forty times, sometimes in water that's knee-deep. I had backpacked the canyon before and had found the experience of hiking in cold, water-logged boots to be less than awesome. I figured that a pair of strong sandals would do the trick—open to allow for quick drying, and sturdy enough to handle the terrain. Turned out to be a bad idea.

This was my second trek in Aravaipa Canyon. I had returned there to begin crafting a personal ritual of pilgrimage: an annual solo trip to the wild, on the eve of the new year, to take stock of things, to mark and measure the progress of my life's path. The desert is an obvious choice for such explorations. As the first prophets knew, the desert enforces clarity. There's just the sun, the sky, and you.

Beyond the archetypes, I had a sentimental reason for heading alone into the desert. I grew up in Arizona, and for me a trip to the Sonoran Desert always feels like going home, the oily scent of creosote like the return to some original state of being. Aravaipa is an especially great place for a vision quest because it's usually empty. The canyon is about 125 miles southeast of central Phoenix, a scant two-and-a-half-hour drive from the maze of freeways and tract homes. Yet I've never seen more than a handful of people there. It's the sort of place where one can experiment with solitude, can imagine the world as newly born.

Aravaipa Creek is a rarity in the desert—a spring-fed creek that flows year-round—and through millennia the water has cut a deep gash into the Galiuro Mountains. The canyon begins with heavy slabs of dark-red shale at the bottom, rises into rust-colored schist, and then rises further into cliffs of orange-and-peach limestone. Eons of the planet's story are visible in a glance, whole epochs etched in the span of a thousand vertical feet.

The canyon slopes are pure Sonora Desert: tall, multi-armed saguaros, writhing agave, prickly pear, and patches of gray bursage and brittlebush. It's a world of heat and thorn and rock. A whole other universe exists just below. Along the creek grow thickets of willow skirted with horsetail reed and cattails. Colonnades of cottonwoods arch above the streambed, where cool green algae cloaks the rocks in the water.

The oasis is home to all kinds of critters. During trips to the canyon I've spotted mallard ducks and green-winged teals and flocks of northern pintails with their long, brown faces. Several times I've scared up a great blue heron, which will flap its wide wings and retreat upstream ahead of me until I surprise it again, and then again. There are whitetail deer and packs of javelina, fierce-looking with their porcupine-like hairs. Once, walking at dusk, I came across a ringtail cat in the grasses near the water. A column of white and black fur threading the reeds, nothing more. The animal was like an apparition, like the remembered half-image of a dream. Aravaipa Canyon is a place full of such marvels.

I was probably in the midst of some such romantic reverie, walking through a grove of ivory-white sycamores and not paying close attention, when I swung my left foot into the jagged edge of a log. The pain was unforgiving, like an ice pick through the flesh. I cursed, stumbled, and almost fell over before I managed to fold myself to the ground to inspect what the hell had happened.

It looked like someone had stapled a busted tree stump to my foot. Long pieces of splintered wood were sticking out of me, and my skin was scratched and torn. The slivers had managed to slide straight into the top of my foot, slicing between the skin and the complex of muscle and bone below. Blood ran through my toes. The situation was bad.

Unfortunately, my intrepidness—a younger man's cavalier courage—outstripped my preparedness. I didn't have a first-aid kit on me. I didn't even have any tweezers. But I did have my Leatherman, and soon I was using the pliers to try to pull the wood out of my foot.

The first and second splinters came out easily enough, and I sighed with relief. The third one took some doing—it was jammed in there pretty good—but I eventually worried it lose. I was working on the last one when the wood split off. A long, flat piece of tree remained buried in my foot.

I wasn't going any further up the canyon and, with the day waning, I wasn't going back downstream, either. I would have to pitch camp.

Feeling jittery, I went to the creek to pump some water and gather myself. My second bottle was nearly filled when, looking across the water, I spotted a lovely little meadow amid the cottonwoods. It looked peaceful over there, welcoming. As if drawn by a magnet, I limped across the creek to check it out.

Above the first meadow I found a second clearing, just as green, and above that a third, almost as if the land had been terraced by hand. I spotted an overgrown path at the edge of a mesquite thicket and followed that for a short fifteen yards until I stumbled upon one of the coziest and most inviting campsites I have ever known.

In a small rock hollow there was a flat space tucked beneath the canyon walls. A primitive hearthstone anchored the tiny clearing. Imagine a large, oval boulder, buried to its midpoint and split in half, so that one side was perfectly rounded and the other flat, and that flat side stained black with the soot of a thousand fires. Just below the clearing I discovered a smooth stone pool edged with bunch grasses and cattails and fed by a side stream trickling out of the cliff heights. It was so pretty, so calming, like some kind of elfin outpost. I couldn't believe my luck.

I hobbled with my gear across the creek and set up my tent. I cleared the fire pit, gathered some wood, put my stove together. It was only a week after the winter solstice, and night came on quickly there in the canyon bottom. As the first stars appeared, I wondered how long people had been coming to this spot.

A long time, I thought. Then I looked at the flame-scorched patina on the fireplace and got the unshakable sense that it had been a *very, very long time*. Centuries. Maybe millennia. The hair on my arms and neck rose on end. Once this was a peopled wilderness.

The Hohokam and the Salado peoples might once have camped here, and who knows which other tribes in the unrecorded years before that. Almost certainly some members of the Aravaipa band of Apache had slept in that little nook nestled in a Y above two streams. Maybe even Ezkiminzin himself, the last chief of the free Aravaipa, had been here. Maybe he had used the place as a refuge after many of his people were slaughtered at the canyon's mouth by a Tucson mob in the spring of 1871. Maybe there had once been songs sung here.

For the Apache after whom the canyon was named, this thin patch of forest in the middle of the desert had been an earthly paradise, their most beloved place. The canyon served as hunting ground, mesquite pod granary, apothecary, mountain shortcut, stronghold, and sacred space. The place was hearth and altar all at once. Which is to say, a sanctuary.

It may sound mystical, or superstitious, but I slept better that night knowing that my camp in the wild had once been someone's home. I felt comforted by the sense that I was not the first person to have found a haven in the wilderness.

I got out of the canyon okay, though with my badly swollen foot it was a real pain. It took me at least six hours of scrambling and splashing through the creek to make it out of the wild. When I finally spotted the works of man, I knew I would be all right.

First, electrical lines, strung to an abandoned inn at the mouth of the canyon. Next, my car at the trailhead, the rumble of its engine signaling my return to civilization. Then blacktop and the wonder of motorized speed. And then a cascade of the technologies that make our lives easy: a cell phone tower on a ridge, an old guy in a white cowboy hat scraping the desert bare with the blade of a bulldozer, the smokestack of the copper smelter in Hayden, surrounded by ziggurats of mine tailings. The scars on the landscape guaranteed that, now back in what we have come to call "the world," I would be fine, my foot would be fine.

My mishap in Aravaipa Canyon happened a decade ago. I've had many other misadventures in the wild since then and have collected many more campfire tales. But the memory has been on my mind a lot lately as I hear rumors that the wildness of the natural world is on the verge of going extinct.

Academics and policy wags at universities and think tanks declare that wilderness no longer matters, if it ever truly existed at all. We are told that we have entered a new epoch of planetary history: the "Anthropocene," or Age of Man. We are told that the human footprint is too big to leave any place untamed, and that all of Earth is a garden to be cultivated by us. "We must abandon our love affair with the wild . . . for the cold light of the necropsy," writes one political scientist, who declares that we have come to "the end of the wild."

Appreciation of wilderness is obsolete in the twenty-first century, according to this way of thinking. Wilderness is irrelevant to our efforts, haphazard though they are, to create an ecologically sustainable society. Our numbers have grown too large for solitude, and our technologies have deleted all the blank spots on the map, filling them in with grid lines. We now live, supposedly, in a "post-wild world."

I don't know what to think about such bold proclamations. They leave me feeling unsettled, confused. What about the wildness there in Aravaipa Canyon? Wasn't that real? Isn't it still important?

My head understands the announcements of the wild's passing. Wildness does seem an endangered species, or at least a cornered beast, like a mountain lion hounded into a tree. And wilderness, as a place, is a scarce and dwindling resource. The efforts to preserve primitive places are, quite literally, losing ground. For the vast majority of people living in urban and suburban America, the wild is nothing more than legend or myth.

But my heart rebels against the idea of a world without wildness. Is it true that we live on a completely tamed planet? *Really?* It's a frightening notion, this vision of a completely gardened Earth. The thought of a "post-wild" world leaves me sad and depressed.

Sure, a world without the wild would have fewer risks and dangers; it would be, by definition, more manageable. But it would make the planet a smaller place, with less beauty and far less magic. In wildness resides mystery—and we need mystery in our lives like we need our daily bread. Mystery nourishes imagination; it is hope's fuel.

Seems to me (to crib a line from Mark Twain) that the reports of the demise of wildness have been greatly exaggerated. I hear the claims about wild's death and they appear to be not so much descriptions of fact as expressions of desire—the ancient human wish for convenience and comfort and control. I think of that crazy trip to Aravaipa Canyon and I wonder whether those who have written wilderness's obituary do not recognize the wild that remains simply because they have not searched for it. The bright lights of our big cities are full of certainties, and wildness is nothing if not a shapeshifter, hard to pin down. Wildness is as sneaky as a coyote: you have to be willing to track it to understand the least thing about it.

But maybe I'm wrong. Maybe wildness has become an anachronism in this Human Age, a mere figment of our imaginations. Maybe the last vestiges of wildness are, in fact, about to be snuffed out. If so, what then?

To try to answer that question, I have spent the last few years going deep into the wild—both the wild of the natural world and the wildness within. I've trekked through wildlands. I've talked with wild men and wild women and tracked wildlife. I've eaten wild foods and slept near wild waters. I've gone to the ends of the earth.

What I found surprised me—and I think it will surprise you, too.

Oh, and what about the big splinter in my foot? It's still there.

When I got back to Phoenix I went straight to the home of a high school buddy whose fiancé was an MD. She took one look at my massively swollen foot, flinched, and called in an antibiotic prescription to stem the infection. By the time I returned to my home in California, the swelling had gone down, but the splinter was still pinching me horribly. After some x-rays and close examinations, the doctors decided not to try to extract the wood. Too many bones in the foot, they said, and besides, it looked to be healing.

That shard of the world remains stuck beneath my skin. When I touch the top of my foot, I can feel it. The splinter has shrunk by now, much of it absorbed by my body, but it's apparent—a nubbin of wood and scar tissue, like a pearl.

I like having it there. It's a reminder of how the wild has made an indelible mark on me.

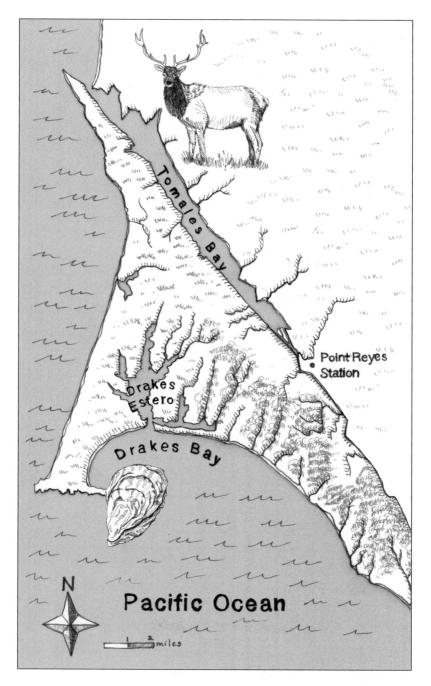

Tomales Bay

Point Reyes
Station

Drakes
Estero

Drakes Bay

Pacific Ocean

N

1 2 miles

Point Reyes National Seashore

1

Bewildered

ONE OF THE STRANGER POLITICAL CONTROVERSIES of the last decade centered on a little creature whose anus runs through its heart: *Crassotrea jurgas*, the common oyster.

The fight over Drakes Bay Oyster Farm had all of the plot points you might expect in a good ol' environmental battle. For starters, a beautiful place—Point Reyes National Seashore, a national park not far from San Francisco that, with its rugged cliffs and stormy beaches, is a postcard for the Northern California coast. Second, charismatic wildlife. In this case harbor seals, whose attitude toward the oyster operation was a matter of heated debate. There were also a mind-boggling number of scientific reports, spiked with accusations of flawed evidence and rigged results. Plus the usual bare-knuckle tactics of politics: Capitol Hill maneuverings, copious media spin, character assassinations, legal appeal after legal appeal, and reams of emotional polemics badly disguised as reasoned arguments. And all of it, naturally, wrapped within big claims about what would be in the public interest—to allow the oyster farm to keep operating, or to create the first large marine wilderness area on the West Coast?

What made the Drakes Bay Oyster Farm controversy so weird was that everyone, on all sides, proudly claimed to be fighting for the environment.

Point Reyes National Seashore lies on the far western edge of Marin County. West Marin is a land of rolling hills stitched with creek-bottom redwood groves and ridge tops of live oak and pine. Dairies and cattle ranches—nearly all of them organic—dot the scenery, and the older families, the ones that have been there for generations, are mostly involved in agriculture or fishing. The rest of the local economy is geared toward serving the tourists—millionaire millennials up from Silicon Valley, or the hordes of Lance Armstrong wannabes who pack the country roads on Saturdays. A lot of artists and writers live in the area, many of them back-to-the-land types who settled there in the sixties or seventies. People in West Marin like their food local and chemical-free, they donate to the community radio station, they think of themselves as big-hearted and open-minded. Most everyone does yoga.

So it was something of a shock to the area's social ecosystem when, in 2005, the fate of the oyster farm began to tear the community apart. Since the 1970s, the oyster operation on Drake's Estero had been managed (quite badly, most locals agreed) as Johnson's Oyster Farm. Then Kevin Lunny, the scion of a longtime ranching family, bought out Johnson, poured a bunch of money into the place, rebranded it as Drakes Bay Oyster Company, and announced his intention to stay as long as possible. That's when the trouble started.

In 1972, Johnson had sold his property to the National Park Service, which gave him a forty-year lease to continue operating. In 1976, the US Congress designated the estuary at the heart of the seashore as a "potential wilderness area"—meaning that when the lease expired in 2012, the estuary would receive full federal wilderness protection. Kevin Lunny's announcement that he wanted to keep raising oysters in the middle of the national park beyond the lease expiration threw the plan into doubt. The battle lines were drawn: Should the oyster farm stay, or should it go?

Folks in West Marin are an opinionated bunch, and soon enough debates about the oyster farm dominated local conversations. It was all but impossible to go into The Western, the saloon in Point Reyes Station, and not hear talk about the oyster farm. Red-hot exchanges erupted in the pages of the *Point Reyes Light* and the *Marin Independent Journal*.

At first, each side made the predictable appeals to science. "Science can be wrong and should be subjected to rigid peer review, but it is never irrelevant," wrote one oyster farm defender. "Those who seek to make it so, or, worse, attempt to suppress it from the record, are either losing a battle where science is proving them wrong, or they are simply intellectual cowards unwilling to sit down and deliberate with someone who has probably studied the situation more." An oyster farm opponent countered: "There are a few scientists who claim oysters are needed for the Drakes Bay ecosystem to function, essentially stating that the ecosystem wouldn't or couldn't function in a pristine state without human intervention. I would respectfully recommend that these individuals review a college-level ecology textbook to see the flaws in their claims." The Oracle of Science is nothing if not cryptic; each side could pick their preferred studies and read them as they liked.

The situation turned nasty. There were accusations of "Taliban-style zealotry." Neighbors stopped greeting each other at the post office. People were disinvited from birthday parties. "The viciousness is beyond anything I have experienced in our community," a reader of the *Point Reyes Light* complained. "The personal attacks, the politics of personal destruction, the career-ending attacks on scientists—frankly, it's disgraceful," Amy Trainer, an oyster farm opponent (or wilderness proponent, take your pick), told me.

Those, like Trainer, who were opposed to renewing the oyster farm's lease made a classic preservationist argument: the estuary is a special place that deserves the highest protections. Also, they said, a deal is a deal. The National Park Service had given the oyster farm a good forty years to stay in business and now, under the terms of the agreement, it was time to restore the estuary to a condition resembling how it had looked for millennia. Besides, the oyster opponents said, the aquaculture wasn't really all that sustainable. Not with its plastic tubing for cultivating nonnative oysters and clams, not with its motorboats disturbing birds and beasts.

A lot of the oyster farm's backers had solid environmental credentials themselves—people like farm-to-table pioneer Alice Waters and Peter Gleick, a leading climate change scientist. They saw the situation differently. To them, the operation represented the ideal of the green economy. Here was a local business, growing local food, and in a way that had a relatively small ecological footprint. Those harbor seals supposedly so inconvenienced by the oyster farm? The farm's defenders liked to point out that they're called *harbor* seals for a reason—they don't mind a bit of human presence. According to its defenders, the oyster farm was an example of how humans could coexist with wild nature. We could have our wilderness and eat it, too.

"There are some parts of Point Reyes Seashore that are considered wilderness, and appropriately so," Phyllis Faber, a vocal oyster farm defender and longtime Marin resident who has been active in environmental causes for more than forty years, told me. "Some parts of the environmental community, they want the whole place to be wilderness. It's an impossible yearning."

We were sitting in Faber's home, a townhouse in a retirement community that's perched near a wetland called Pickleweed Inlet, one of the San Francisco Bay's hundreds of small fingers. Faber, white-haired and energetic, served me tea. Outside the window I could spot scores of birds puttering about in the saltwater marsh. "I don't think they know what they're talking about," she said. "The environmental community, it's wishful thinking on their part, to think this is wilderness. They have a fantasy of what they would like. It isn't very realistic."

This from a woman who was one of the first members of California's Coastal Commission, a person who describes herself as a great bird lover, who for years edited *Fremontia*, the magazine of the California Native Plant Society. For such an ardent nature-lover, like many other nature-lovers, to come out against wilderness protection—well, it was bewildering.

As parochial as it seemed at times—a battle royal over bivalves!—the oyster farm controversy had cast into sharp relief some of the most difficult questions about our relationship with Earth. What do we expect from wild nature? Wilderness on a pedestal? Lands that we garden and tend? Or something in between?

Does wilderness have to mean "pristine"? How can we include history and memory in our idea of wilderness? Where do we draw the line between human actions that are beneficial and those that are harmful?

And the biggest question of all: with the human insignia everywhere, is there any place or any thing remaining that is really, truly *wild*?

Full disclosure: Point Reyes National Seashore is one of my favorite places in the whole world. I've lived in the San Francisco Bay Area for close to twenty years, and during that time I've explored all the corners of the park. Dozens of times I've climbed up and over Inverness Ridge, where thick forests of Douglas fir trees catch the morning fog to make their own rainwater. I've hiked the aptly named Muddy Hollow trail in the middle of winter, when the paths are thick and soggy. I've covered all sides of the estuary at the center of the park, counting the dunlins in the mudflats and the loons in the shallows. One of my favorite spots is Abbot's Lagoon. There's a touch of everything in the scene: a freshwater pond fringed with tule reeds; a saltwater estuary usually busy with shorebirds; pastureland; and the smash of the surf just beyond the sand dunes.

Point Reyes is a triangle-shaped peninsula jutting into the Pacific Ocean that, from the air, looks as if the coastline is giving a giant *hang loose* sign. The pinkie tip at the north end is Tomales Point, the thumb at the south end is Chimney Rock, and in between is a fifteen-mile-long knuckled stretch of beaches and cliffs. A large, claw-shaped estuary lies in the middle. This is Drake's Estero, named after the English swashbuckler Sir Francis Drake, who, in the summer of 1579, beached his ship, the *Golden Hind*, there for repairs.

"A faire and good Baye," Drake called the place, which he christened Nova Albion—"New England." The tall white bluffs above what is now called Limantour Beach do, in a way, resemble the Cliffs of Dover. The peninsula's interior—fog-shrouded, all but treeless—also has a certain English vibe. Arriving as he did in the dense summer fog, it was easy for Drake not to have spotted the opening of the Golden Gate, just miles away. Whether the native people Drake encountered and traded with, the Miwok, tried to tell him of the vast bay to the south is unclear. In any case, he just barely missed "discovering" one of the greatest natural harbors on the planet.

Today, one of the most amazing things about Point Reyes is its proximity to civilization. It can take as little as an hour to get from the middle of San Francisco to a trailhead. In the long light of summer I can spend a full day in the city, cross the Golden Gate Bridge at five o'clock, be hiking through the shoreline grasses by six, and arrive at Coast Camp at dusk, where, if I am lucky and the visibility is clear, I can watch Venus emerge over the silhouette of the Farallon Islands. It's a journey into another world, made in the space of an evening.

This closeness to civilization—which now seems a virtue—was once a liability. In the boom years after the Second World War, the suburbs of San Francisco began to encroach on what had long been a farming community. Real estate interests were carving roads for subdivisions into the headland overlooking Limantour Beach. Logging was under way amid the moss-cloaked fir and bishop pine of Inverness Ridge. People feared the peninsula's charms would be lost to development.

Conservationists, led by the Sierra Club's David Brower, launched a campaign to stop the bulldozers. They won. In 1962, Congress passed and President Kennedy signed a bill "to save and preserve, for purposes of public recreation, benefit, and inspiration, a portion of the diminishing seashore of the United States that remains undeveloped."

The creation of Point Reyes National Seashore was one part of a larger burst of conservation activity that occurred in the mid-twentieth century. In the years following the Second World War, many Americans were starting to feel an uncomfortable sense of being hemmed in. The farms and fields they had grown up with were turning into sprawling suburbs. The country's population had just surpassed 100 million, making some people feel they were being overcrowded. The automobile was reshaping the country as the Interstate Highway System shrank distance, making once-inaccessible places all too close. Because of "the brutalizing pressure of metropolitan civilization," one group of conservationists declared, it was essential to keep some lands "primeval" and "virgin . . . free from the sights and sounds of mechanization." By the 1950s, an increasingly well-organized and focused citizens' movement was demanding that the remaining wild places be preserved.

The result was a string of conservation victories unprecedented since the presidency of Theodore Roosevelt and unmatched to this day. In the space of a decade, the Sierra Club's Brower—assisted by powerful allies such as Secretary of the Interior Stewart Udall and Lady Bird Johnson, the First Lady—protected millions of acres of land. During Udall's tenure, the National Park Service expanded to include Canyonlands National Park in Utah, North Cascades National Park in Washington, California's Redwood National Park, and Cape Cod National Seashore, plus six national monuments, nine recreation areas, and fifty-six wildlife refuges. Congress passed the Wild and Scenic Rivers Act and the National Trails System Act, which formalized the Appalachian Trail that stretches from Georgia to Maine.

One of the major achievements of the time was the passage, in 1964, of the Wilderness Act, which ranks among the signature accomplishments of the American environmental movement. Howard Zanisher, then-president of The Wilderness Society, wrote the opening sections of the Wilderness Act. The legislation is remarkable for a quiet poetry that is so rare in lawmaking. The act says, "A wilderness, in contrast with those areas where man and his own works dominate the landscape, is hereby recognized as an area where the earth and its community of life are untrammeled by man, where man himself is a visitor who does not remain."

Those words represent a fundamentally radical and history-breaking change of mind. Here was a nation founded upon an antagonism against the wild—the English Puritans at Moment One declaring a war against "the howling wilderness"—that had come to revere wilderness as something fundamental to its character. A country that by 1964 had gone so far as to codify in law a definition of wilderness as a place that would not be subjugated by human will. The pivot from fearing the wilderness to loving the wild is one of Americans' most important contributions to human thought. Like the national parks system that preceded it (famously, "America's Best Idea"), the Wilderness Act reversed centuries of thinking regarding how humans are supposed to treat the rest of creation. It was a truly original idea, this notion that the land might have its own interests apart from ours.

But the Wilderness Act went much further than the national park ideal. To many mid-twentieth-century conservationists, the creation of national parks wasn't enough to protect the essential character of wilderness. In the fifties and sixties, the park service was dominated by a kind of Disneyland mentality. "Industrial tourism" is how Edward Abbey described what he saw happening in Utah's red-rock country. To park officials of the time, a paved path was better than a dirt trail, a fully equipped cafeteria preferable to a backcountry hut. The parks were being designed mostly to make everything automobile accessible. Still, some people wanted something different. They wanted a guarantee that a few places would be permanently protected from the engine and the asphalt.

One of the most eloquent appeals for the Wilderness Act came from Wallace Stegner, the Pulitzer Prize–winning author. In a 1960 letter to a government official (later published in the *Washington Post*), Stegner made a forceful case for wilderness as a spiritual tonic, a psychological retreat, and a civic good. "Something will have gone out of us as a people if we ever let the remaining wilderness be destroyed. The reminder and reassurance that it is still there is good for our spiritual health even if we never once in ten years set foot in it." In a jibe at the technocratic thinking of postwar government and business elites, he argued that simply the *idea* of wilderness—knowing that someplace remains uninhabited and "only a few people every year will go into it"—has a transcendent value. "Being an intangible and spiritual resource, [wilderness] will seem mystical to the practical-minded—but then anything that cannot be moved by a bulldozer is likely to seem mystical to them." In a now oft-quoted line, Stegner concluded that wilderness is "the geography of hope."

At least as measured by acres protected, the Wilderness Act has been a success, far surpassing the original intentions of its framers. Today, about 110 million acres of land in the United States are protected as wilderness. Those 110 million acres account for about 5 percent of the total US landmass; when you factor out the vast wilderness areas of Alaska, about 2 percent of the territory of the Lower 48 is protected as wilderness.

The Wilderness Act preserves a good-sized chunk of Point Reyes National Seashore. About 30,000 acres of the seashore's 71,000 acres are designated as the Phillip Burton Wilderness, which means they are protected from road building and other permanent infrastructure. Unlike most other national parks, however, the remainder of the seashore—the front country—isn't just for sightseeing. Much of Point Reyes is what's called a "working landscape." That is, it's farmland.

When the national seashore was proposed, many local ranchers and dairymen were vehemently opposed. Their families had been there since the Gold Rush, and they had no wish to leave. So Congress crafted a compromise. The federal government would buy out the farmers and then lease the land back to them so they could continue their agricultural traditions. The land would be both protected and productive. Conservationists were happy enough with the deal. Pastureland, they figured, was preferable to subdivisions.

The shared arrangement continues today. The 2 million people who visit Point Reyes every year find an eclectic mix of working dairies and ranches, undeveloped beaches, and steep forests marked only by footpaths. Point Reyes is home to several different natures: the pastoral nature of the ranchlands, the wild nature of the woods and marshes, and the ecotone where the two meet. In Point Reyes, the untamed and the domesticated overlap. Coyotes thread their way through dairy pastures, and tule elk graze the same hills as cattle. Bobcats are common. One time, hiking the Glenbrook Trail above the estero, I came upon a deer leg, fresh and half-eaten, and I knew that the seashore is a place where the old laws of fang and claw still rule.

But the coexistence between the domesticated and the untamed isn't easy. As the oyster farm controversy showed, our ideas of what we expect from the pastoral and what we hope for from the wilderness are often at odds.

Perhaps a certain friction is inherent to that landscape. Point Reyes straddles the San Andreas Fault, and the park's placid scenery belies a massive tension below. On the east side of the park, the fifteen-mile-long finger of Tomales Bay—formed by an earthquake long ago—traces the line where the Pacific and North American tectonic plates meet. Or *clash*, you could say.

The oyster farm battle was another clash of place, evidence of a rift among people of supposedly similar values. A fissure had opened beneath the ideal of wildness.

We met at Drake's Beach just after sunrise. Low tide was the only time that we would be able to follow the western shoreline to the mouth of the estero, Amy Trainer had said. In 2010, Trainer was hired as the executive director of the West Marin Environmental Action Committee—a local group that has been around since the seventies—and she quickly became one of the most vocal proponents for extending full wilderness protection to the estuary. I wanted to understand the controversy from her point of view, and she had agreed to meet me, suggesting the estero itself. As we skirted the surf, careful to keep a good distance from the elephant seals lazing on the sand, Trainer shared some of her backstory.

She grew up in Kansas and spent many summers in the mountains of Colorado, where she "fell in love with the natural world." At sixteen she "discovered the environmental movement and was, like,

'This is what I want to do with my life.'" She became a vegetarian. She named her dog Henry in honor of Thoreau. At the University of Kansas she spearheaded a campaign to prevent a road from being built through a wetland. She went to law school, where she specialized in environmental law, and then held a series of jobs at regional environmental groups in Washington and Colorado.

Trainer had been at her new job in West Marin for three weeks when she concluded that the law and science called for removing the oyster farm. "I think all of the industrial-scale disruptions—the pressure-treated wood racks, the motorboats frightening the animals, especially migrating birds—it isn't okay in a national park, much less in a wilderness area." The oysters, she explained, "are a nonnative species, a monoculture in an otherwise ecologically pristine area." She admitted that what she called the "attacks" against her for her views had been tough. But she tried to wear the criticism as a badge of honor. "There are a lot of haters, but there are also a lot of people who love wilderness."

We trailed the edge of Horseshoe Pond, zigzagged through a boggy spot filled with juncus, and climbed the bluff above the beach. Patches of purple Douglas iris dotted the grassy slopes, just now turning green in what had been a season of drought. To the east, the sun climbed above the trees on Inverness Ridge. Trainer led me up and down the hills, past a small seep trickling into a no-name pond, and then back up again until, after some bushwhacking through the coyote brush, we found a little perch overlooking the estuary.

I asked the most obvious question I could think of, trying to get at something elemental: What, exactly, was she trying to protect?

"It gets back to the rights of nature," Trainer said. "Do you believe we should incorporate ourselves into every ecosystem? Or are some places sacred and special, places like this? Being able to come out here without the signs of this private operation, without man's fingerprints, to see it existing as it did for millennia. A lot of people don't see the value of that these days. But especially in this age of telecommunications, when we're always being pulled out of ourselves, this is a place where you can pull yourself inward. This is a sacred place. It's a church for many people. I think that's more important than ever in this day and age. It's so valuable on so many levels. You really get this sense of being lost. And also this sense of being connected. This sense of awe and reverence just flows out of you."

I could see what she meant. The brassy light of the early morning sparkled on the estero's green waters. First a pair of white egrets and then a trio of great blue herons glided by, headed for a rookery in a nearby pine. Dozens of harbor seals dozed on the sand bars, their occasional yawps the only sound besides the steady roar of the breakers on Limantour Beach. To the north I could see the oyster racks in the water—straight lines etched into the natural contours, made obvious by the ebb tide.

But most of Trainer's neighbors were unconvinced. She was as eloquent a defender of the wild as you could imagine, her intellectual clarity and self-confidence fueled by a deep passion. Yet even among the environmentally minded folks of Marin County her arguments had failed to persuade.

It wasn't that the Marin locals were against the principle of wilderness protection. "What right-minded environmentalist can argue with the sacrosanct idea of preserving wilderness?" an area resident wrote in the *Point Reyes Light* at the height of the controversy. Rather, they weren't sure that the estero fit with their image of what constituted a wilderness. The letter writer continued: "Leaving aside the obvious point that Drake's Estero isn't a wilderness area, there is another question. Exactly what difference will it make to the environment if Drake's Estero is designated as a wilderness area?"

I heard similar doubts when I went to talk to Phyllis Faber. As I mentioned, Faber has an environmental CV to rival Trainer's. In 1970, when she was a young woman fresh out of Yale and teaching high school biology, she organized a giant celebration on the first Earth Day, the memory of which still electrifies her. Then she and her husband moved to California, and Faber jumped into environmental activism. She helped spearhead the effort to pass a state initiative, Proposition 20, to put in place the nation's toughest coastal protection law. With an area dairywoman she cofounded the Marin Agricultural Land Trust to protect open space from housing developers. For years she worked for the University of California Press, editing books about native flora. (My copy of *Designing California Native Plant Gardens*, which she edited, is dog-eared from years of reference.)

Like Amy Trainer, she was dismayed by how personal the dispute over the oyster farm had become. "I am really at odds with the environmental community, which is really unfortunate, because I've been a part of that community since the sixties," she told me. And, like Trainer, she was fired up. "It's patently stupid to want to get rid of the oyster farm. I am just enraged by this."

I asked Phyllis Faber the same question I had posed to Amy Trainer: What, exactly, was she trying to protect? "I'm trying to protect the ecosystem in Point Reyes. I'm trying to protect the coastal law. I'm trying to protect our community. The oyster farm doesn't damage the marine preserve—the oyster farm *benefits* the marine preserve. It's the kayakers that disturb the harbor seals."

She paused, took a sip of her tea, and then said, "It's a great benefit for people to come out and see a place that doesn't have houses all over the place, that doesn't have the signs of man. I think that's wonderful. But wilderness is different. It implies the *absence* of man. Their [the oyster farm opponents'] notion of wilderness is unrealistic. That land hasn't been wilderness for 200 years. It had Indians on it for centuries, and European farmers on it for more than a hundred years. There are roads. There are farms."

By the time Interior Secretary Ken Salazar announced, in November 2012, that the park service would not renew the oyster farm's lease, Faber's view had become conventional wisdom in West Marin, where scores of hand-painted "Save Our Drakes Bay Oyster Farm" signs dotted the roadsides. "There is no ecosystem in Marin that has not suffered from the influence of humanity," an area fisherman wrote in the *Light*. "Humanity is constantly meddling with nature." Another reader argued, "I find wilderness an unusual concept in a park located within an hour or so of 7,000,000 people and on which there is an extensive road system, parking lots, and sanitary facilities."

A local architect, a guy sometimes referred to as the grandfather of ecological design, put the case to me most succinctly: "Wilderness is a fantasy."

At the heart of almost every environmental battle lies the question of where we think humans fit in the natural world. Does the whole planet exist for our benefit, to be cultivated by us like a garden? Or do we have a responsibility, a moral obligation even, to leave untamed as much of the world as possible? Or is it a little of both?

Many people have always objected to the idea of wilderness, the notion that we would keep some places off-limits from human appetites and "lock them up." But, at least among self-described environmentalists, wilderness preservation has long been a bedrock principle, the animating force of more than a century of conservation efforts. As the fight over the oyster farm revealed, those once-solid beliefs are looking shaky.

This is not because the folks of West Marin have a callous disregard for wild nature. The oyster controversy had no villain from Central Casting calling for more tar-sands oil extraction or wanting to blow the top off a mountain to get at coal deposits. Point Reyes Station has been called "the greenest town in America." Bolinas, a village at the south end of the seashore, has a sign at its entrance declaring that it is a "socially acknowledged nature-loving town" (whatever that means).

Rather, the shift in thinking is a response to the new realities of the twenty-first century. On this overheated and overcrowded planet, priorities are changing. The old faith in the value of wildness is melting under the glare of a hot, new sky. For many people—including those who would call themselves environmentalists—human self-preservation now trumps the preservation of wild nature. The love of the wild may be, to borrow a biologist's term, *maladapted* to the new age that some are calling the Anthropocene.

The Anthropocene—the Age of Man, or, simply, the Human Age. If you haven't encountered the word much yet, you soon will. The term was coined in 2000 by the Nobel Prize–winning chemist Paul Crutzen (the scientist who discovered the hole in the ozone layer) as a way of describing the fact that human civilization is now the greatest evolutionary force on the planet. "It's we who decide what nature is and what it will be," Crutzen has written. The neologism is on the verge of becoming scientific standard. In 2008, the Stratigraphy Commission of the Geologic Society of London accepted a proposal to consider making the Anthropocene a formal unit of planetary time. The society's members are now reviewing the idea. By the time this book reaches your hands, it's likely that scientists will have declared that we have, officially, left the Holocene, the epoch in which human civilization was born, and have entered a new period in Earth's history.

It's hard to know what to make of such a big idea. The notion of a planetary age named after humans seems in bad taste, the old colonialist habit of wrecking a place and then putting your name on it. Even when the term is intended as a warning, declaring an epoch in our honor is two parts chutzpah to three parts hubris.

But it's impossible to argue with the facts of our overweening power. Cities and farms dominate the terrestrial landscape. We've remodeled the seas and the sky, too, as our industrial effluent heats the atmosphere and alters the pH of the oceans. With our huge population, we are steadily destroying the habitats of plants and animals and causing a mass extinction not seen on the planet in millions of years. The list goes on and on: our synthetic products have disrupted the planet's chemistry, our lights have blotted out the stars, our accidents have created atomic forests. We've even created a new stone—*plastiglomerate*, formed when plastic melts and fuses with rock fragments, sand, coral, and shells. There is no place and no thing on Earth that humans have not touched.

To accept the idea of the Anthropocene does not necessarily justify the vast alteration of Earth the term describes; it's just to acknowledge the facts as they are. The issue then becomes: What do we do with the fact of the Anthropocene's impending arrival? If all of Earth is ours, where in the world does that leave us?

A growing chorus of writers and thinkers have come to the conclusion that the dawn of the Anthropocene means the sunset of wildness, and that we had best come to terms with our role as the ruler of life on the planet. "Ecosystems will organize around a human motif, the wild will give way to the predictable, the common, the usual," writes the late Stephen M. Meyer in his bleak monograph, *The End of the Wild*. Novelist-naturalist Diane Ackerman takes a more upbeat tone in her book *The Human Age*, yet her cheerfulness seems to just gloss similarly depressing thoughts. "In the Anthropocene," she writes, "it can be hard to say . . . what we mean by a 'natural' environment." At this point, she says, "we must intervene" in wildlands to try to save species from our own threats. "Nature has become too fragmented to just run wild."

One of the most ambitious reevaluations has come from a science writer, Emma Marris, whose *Rambunctious Garden* was one of those books that launched a thousand blog posts. According to Marris, we now live in a "post-wild world." She writes: "We are already running the whole Earth, whether we admit it or not. To run it consciously and effectively, we must admit our role and even embrace it. We must temper our romantic notion of untrammeled wilderness and find room next to it for the more nuanced notion of a global, half-wild rambunctious garden, tended by us."

This rethinking of wildness and wilderness has been a long time coming. In the last fifteen years or so, "no concept has been more hotly contested within the American environmental community than wilderness," according to historian Paul Sutter. If there's any one person who could lay claim to sparking the discussion, it would probably be an environmental historian named William Cronon.

In 1995, Cronon, a professor at the University of Wisconsin, published an essay in the Sunday *New York Times Magazine* with the provocative title, "The Trouble with Wilderness; or, Getting Back to the Wrong Nature." Cronon began by pointing out that wilderness is a human construct, "an entirely cultural invention" formed during the specific conditions of the American frontier of the nineteenth century. "My criticism in this essay is not directed at wild nature per se," Cronon wrote, "or even at the efforts to set aside large tracts of land, but rather at the specific habits of thinking that flow from this complex cultural construction called wilderness." And what were those "habits of thinking" that had made the wilderness ideal "potentially so insidious"?

The first and most worrisome was the way in which wilderness might encourage us to compartmentalize nature as something apart and away from daily human life. "By teaching us to fetishize sublime places and wide open country, these peculiarly American ways of thinking about wilderness encourage us to adopt too high a standard for what counts as 'natural,'" Cronon wrote. Such a view sets wild nature and human culture irrevocably in opposition to each other: "It makes wilderness the locus of an epic struggle between malign civilization and benign nature." And it justified the removal of people from our picture of wilderness—most tragically, the way in which Indigenous peoples, from the Shoshone in Yellowstone to the Ahwahnechee in Yosemite, were kicked off their lands to make way for nature preserves.

The essay upset environmental leaders, who feared the criticisms would undermine public support for conservation efforts. But the critique had already slipped into the intellectual bloodstream. A year before Cronon's essay appeared, a young essayist and anti-nuclear activist named Rebecca Solnit published *Savage Dreams*, a book that devastatingly deconstructed the myth of Yosemite. "By and large Yosemite has been preserved as though it were a painting," Solnit wrote. "Looking is a fine thing to do to pictures, but hardly an adequate way to live in the world." Around the same time, a New York journalist named Michael Pollan published his first book, *Second Nature*. He argued that the garden, rather than the wilderness, might be a better metaphor for thinking about our relationship with the more-than-human. The wilderness ethic, Pollan wrote, "may have taught us how to worship nature, but it didn't tell us how to live with her."

The insights from Cronon, Pollan, and Solnit (among others) could be filed under the category of constructive criticism. Recently, however, the critiques of wilderness sparked by the arrival of the Anthropocene have taken on a sharper, antagonistic edge. "Conservationists will have to jettison their idealized notions of nature, parks, and wilderness," we are told. "A new conservation should seek to enhance those natural systems that benefit the widest number of people." Which is to say, we should give up on the wild, and recognize that the world is ours to improve upon as we see fit.

That argument comes via a trio of biologists—Peter Kareiva, Michelle Marvier, and Robert Lalasz—who argue that "the unmistakable domestication of our planet" means that it's time to dump what they call "conservation's intense nostalgia for wilderness and a past of pristine nature." The biologists' argument appeared in a manifesto titled "Conservation in the Anthropocene." In their essay the three biologists assume that "there is no wilderness," and declare that "protecting biodiversity for its own sake has not worked." Instead, we should embrace the idea that "nature could be a garden . . . a tangle of species and wildness amidst lands used for food production, mineral extraction, and urban life."

"Conservation in the Anthropocene" sparked a heated backlash from other conservation biologists, who objected to its tone (so little lament and so much I-told-you-so) as much as to its content. For a couple of years now, editorials and essays and argumentative rebuttals have flown back and forth in scientific journals such as *Conservation Biology* and *Animal Conservation* as scientists and environmental advocates debate what's more important: protecting wild nature for its own sake or for what it can provide to humans, its "ecosystem services." The squabble has swelled into a public schism, with headline writers declaring a "Battle for the Soul of Conservation Science." The animosity has become so intense that, at the end of 2014, a pair of eminent scientists penned an open letter in the journal *Nature* calling for a détente.

Wilderness has always been contested terrain. It would be possible to dismiss "Conservation in the Anthropocene" as yet another reactionary attack against the idea of "locking up land" were it not for this: the lead author, Peter Kareiva, is the chief scientist of The Nature Conservancy. This is the same Nature Conservancy that describes itself as "the leading conservation organization working around the world," that claims more than a million members, that has as its homepage the enviable URL www.nature.org.

Let that sink in for a minute. According to the head scientist of the world's largest conservation organization, wilderness is dead.

I should make my allegiances clear. I am a pastoralist—that is to say, a gardener. I don't just mean that I'm a gardener in the sense that I grow some kale and green beans in my backyard (though I do that, too). I'm what you could call a Pro-Am Gardener.

In 2005, I cofounded the largest urban farm in San Francisco. At Alemany Farm we grow about 16,000 pounds of organic fruits and vegetables annually on a smidgen of land in the one of the densest cities in the United States. In the summer the farm is packed with rows of tomatoes, strawberries, cucumbers, and squash. In the winter we grow cabbages, chard, and collards. We have a hillside orchard with some 140 fruit trees—apples, pears, plums, avocadoes, mulberries, quince, lemons. We have a perennial stream that flows into a pond ringed with tule reeds, and every spring the pond is busy with huge flocks of red-winged blackbirds and the occasional mallard duck pair. We have beehives. And all around, on every side, is concrete and asphalt. To the east of the farm, next to our greenhouse, is a 165-unit public housing complex. To the south, the constant rush of eight lanes of Highway 280. To the north, condos.

Over the years I put a huge amount of blood, sweat, and tears into building a farm next to a freeway because, in part, I had internalized the critiques of the romantic wilderness. For me, the wilderness revisionism was conventional wisdom, and I was eager to explore a closer kind of nature. I had worked as an environmental journalist for a while, had spent some time living on an organic farm, and I possessed a passion for ecological sustainability and environmental justice. I was determined to try, in my own small way, to ignite a similar passion in others. But I knew that I was unlikely to replicate my own conversion experience of standing among the row crops reading poems in the morning fog. So I figured: if you can't take the people to the land, then bring the land to the people.

I had read my Michael Pollan, I had read my Wendell Berry, and I agreed that the ancient act of (non–chemically intensive) agriculture was as good a way as any to prompt people to recognize our reliance on natural systems. An urban farm could help teach people about how to coexist with our environment. The garden could be a vehicle for getting people to understand that we are "dependent for [our] health and survival on many other forms of life," as Pollan has written. A sun-warmed strawberry, sprung from the dirt—what a wonderful example of how, in Berry's words, "we are subordinate and dependent upon a nature we did not create."

I'm proud to say that it worked. The kids from the housing projects like catching frogs in the creek and getting to pick apples right off the tree. They might not have the same privileges as other San Francisco kids in this dot-com Gilded Age, but they're the only ones in the city who have herons and egrets wading through a pond in their backyard. Techies come out to the farm during our regular community workdays and have their minds blown by the simple fact that we grow food using horseshit. I'll never forget the time I was weeding a bed with a student from San Francisco City College. I asked him why he had come out to the farm. "It's just great to get back to nature," he said. I almost dropped my hoe. *Get back to nature?* Didn't he hear the traffic rushing past?

Or maybe he did, and that's the point. The great virtue of Alemany Farm—like Point Reyes National Seashore—is that it offers experience with what you could call the "nearby nature."

The nearby nature isn't as head-spinning and heart-throbbing as remote wilderness. But what it lacks in intensity it makes up for in intimacy. For many of us, the backyard woodlot or the local beach are the natures we hold closest to our hearts. Let's say you visit the same regional nature preserve every week for years. You come to know the trails in snow as well as in summer. After a while you are so attuned to the space that you can feel how the slightest shifts of weather change its mood. With the accretion of the seasons a place becomes a character. If that character-place is a garden, the relationship may go even deeper to something resembling a partnership.

The wilderness "is a sublime mistress, but an intolerable wife," Ralph Waldo Emerson once warned John Muir in a private letter. Flip the metaphor, and the nearby nature becomes our spouse, the nature we live with, week in and week out. Like any long-term relationship, the rewards are counterbalanced by a litany of frustrations and resentments, the many days that fall so far short of perfection. But if you're being a decent partner, out of that struggle comes a hard-won wisdom about what it takes to balance your desires against those of another. The nearby nature teaches patience and generosity toward the nonhuman.

I strongly believe in the importance of the nearby nature. I believe, as the organic pioneer Alan Chadwick said, that gardening is a formative experience and that "the garden makes the gardener." The garden can force us to recognize our interdependence with nonhuman nature. It can forge an environmental ethic by encouraging us to see that coexisting with wild nature is, in the words of Wendell Berry, "the forever unfinished lifework of our species."

But as I keep hearing the new critiques of wilderness, I fear that they have gone too far. The criticisms are true. But they're only half true—not incorrect, just incomplete. It seems to me that the harsh critics of wilderness have taken an epistemological difficulty (we can't know what an "original," nonhuman nature looked like) and inflated it to a phony existential dilemma (best to dispense with wilderness as a thing of value). There's a technical term for this: throwing the baby out with the bath water.

The more conscientious critics of wilderness have argued that the wild is essential, but insufficient, for creating an ethic of responsibility toward Earth. It seems that the same could be said of the mindful garden: essential, but insufficient, for cultivating an appreciation of wildness. And a deeper and stronger appreciation for wildness is exactly what we'll need to navigate the Anthropocene.

If all of Earth is now our garden, then the garden metaphor has reached the limits of its usefulness—at least as an idea that can blow your mind about what it means to live in harmony with the rest of creation. We require stronger medicine. We need something that can shake us out of ourselves—and wild things accomplish that like little else. A renewed respect for the wild can check the delusion that somehow it has become humans' responsibility to take control of everything and every place.

As a public gardener of sorts, I'm a committed pastoralist. I'm also a passionate backpacker. Over the years I've tramped the middle sections of the Appalachian Trail; the mountains of California, Washington, Montana, and Alberta; the deserts of Arizona and New Mexico; the forests and fjords of Alaska; the Patagonian steppe. While I've learned a great deal from my urban farm, I have also long appreciated the poet Gary Snyder's line that "the wilderness can be a ferocious teacher."

The wild offers different lessons from those of the garden. I'll agree that wilderness—as a "place where man himself is a visitor"—does a poor job of instructing us *how* exactly to live day to day with nonhuman nature. But wilderness provides something just as fundamental: it supplies the reasons *why* we should try to coexist with nature. The wild inspires us to make the effort in the first place.

The more I hear talk about the epoch of the Anthropocene and the arrival of a "post-wild" world, the more I worry we're about to lose that inspiration.

Wild.

I looked it up. I went to my two-volume Oxford English Dictionary. I got out the magnifying glass for reading the miniscule text.

The first definition read: "Of an animal: Living in a state of nature; not tame, not domesticated." The word comes from the Old English *wildedéor*, or wild deer—the beast in the woods. Go further back into the etymology and the meaning becomes more interesting. In Old Norse, a cousin of Old English, the word was *villr*. "Whence WILL," my OED says, meaning that *wild* shares the same root as *willfulness*, or the state of being self-willed. A description lower down the page makes the point plain: "Not under, or submitting to, control or restraint; taking, or disposed to take, one's own way; uncontrolled. . . . Acting or moving freely without restraint."

Notice that there is nothing about being "unaffected" or "untouched"—words that have more to do with the pristine than with the wild. Rather, the meaning centers on the word "uncontrolled." To be wild is to be autonomous, with the power to govern oneself. The wild animal and the wild plant both rebel against any efforts at domestication or cultivation. Yes, we might hunt the *wildedéor*; we might even kill it. But the *wildedéor*'s last act will have been to run free.

If wild is a quality of being, then wilderness is that place where wildness can express itself most fully. You can think of wilderness as "self-willed land." Wilderness is any territory not governed by humans, a landscape where the flora, fauna, and water move "freely without restraint." Wilderness is a place where human desires don't call the shots.

This is a well-trodden path. An appreciation of the untamed is one of the founding principles of environmental philosophy. The wild—as a place and as a state of mind—is as close as you can get to the triggering ideal of environmentalism. For a century and a half, the wild has served as the bright through-line of efforts to preserve the world in something approximating its pre-civilization condition.

Such thinking began, as you might have guessed, with Henry David Thoreau. In an essay titled "Walking," Thoreau dives into a meditation about the meaning of the wild and declares that "all good things are wild and free." Eventually, after a couple thousand words, he works himself up to this now-famous line: "What I have been preparing to say is, that in Wildness is the preservation of the World."

Author-activist Bill McKibben calls the sentence "one of the great koans of American literature." Indeed, the line both requires and resists explication—kind of like the wild itself. If anything, the elusiveness of Thoreau's meaning has only made the call of the wild more irresistible. Inspiration doesn't necessarily require clarity; we are attracted to wildness precisely because it remains always just beyond our reach.

Since Thoreau, the wild has inspired poets, philosophers, and rebels. Wildness has formed the basis of environmental ethics: "The love of wilderness is . . . an expression of loyalty to the earth," proto-monkey-wrencher Edward Abbey wrote. Wildness has been praised as a psychological tonic, an antidote to the confines of civilization: "The most vital beings . . . hang out at the edge of wildness," Jack Turner, a philosopher, has written. And wildness has been celebrated as a civic virtue, an essential ingredient of political liberty: "The lessons we learn from the wild become the etiquette of freedom," poet Gary Snyder writes. Wildness is the heartbeat of a worldview.

There's no question that this North Star has been dimmed. The official preserves of our American wilderness system can feel awfully tame. At the trailheads, signs from the U.S. Forest Service or National Park Service sometimes warn, in a nervous-aunt tone, that falling trees and rocks can cause injury or death. In most places, the paths are marked by cairns to make sure you don't get lost. It is illegal—indeed, punishable by a fine—to sleep in the backcountry of our national parks without a permit. The wildlife is also carefully managed. Federal biologists implant wolves and grizzlies with ID microchips and place GPS collars around their necks, equipment sophisticated enough so that a technician hundreds of miles away can tell whether a bear is sleeping or screwing. Even the animals, it seems, are stuck in the matrix.

I've only lived in a fallen world, and I take it as a given that every place and every thing has been touched by civilization. In my lifetime, humans have destroyed half of the world's wildlife as our own numbers have doubled. By the time I was born, satellites had already embellished the firmament, the radioisotopes of nuclear tests were already scattered in the geologic record, toxic chemicals had already drifted to the North Pole. So I assume there is no pristine nature. I accept that we live in what you might call a "post-natural world."

It is much more useful, then, not to ask what is natural, but to seek out what is *wild*. Because even in its diminished state, the wild still holds a tremendous power. When we search out the wild, we come to see that there is a world of difference between *affecting* something and *controlling* it. And in that difference—which is the difference between accident and intention—resides our best chance of learning how to live with grace on this planet.

In short, what I have been preparing to say is this: it's time to double down on wildness as a touchstone for our relationship with the rest of life on Earth.

If, in the Anthropocene, nothing remains that is totally natural, then the value of wild animals and wild lands becomes greater, if for no other reason than that those self-willed beings remain Other than us. And we need the Other. As a species we need an Other for some of the same reasons that, as individuals, we have other humans in our lives. They center us. By opposing humans' instincts for control, wild things put our desires in perspective. Peter Kahn, a pioneer in the field of eco-psychology, writes that wild animals "check our hubris by power of their own volition." In much the same way, wilderness—or any self-willed land—can remind us that the rest of the world doesn't exist in relation to us, but that we exist in relationship to other beings.

The idea that every landscape should be a vehicle for our desires is species narcissism on a planetary scale. When all of Earth is our garden, then the world will have become like a hall of mirrors. Each ecosystem will contain some glimpse of our own reflection, and we'll be everywhere, with nothing to anchor us. We'll be lost.

A "post-wild" world would put human civilization into a kind of solitary confinement. There would be no Away, no frontier or edge to civilization. There would be no Other, nothing to contest our will. We would be left all alone.

Do you know what happens to people who are placed in solitary confinement? They often go insane.

The northernmost edge of Point Reyes National Seashore is a narrow, ten-mile-long peninsula called Tomales Point. I've hiked it many times, and over the years I've come to think of it as a place that represents a lot of what is best, and much that is troubling, about the twenty-first-century wild.

Tomales Point is protected as a federally designated wilderness area—11,000 acres of shoreline and tall cliffs. Coyote brush covers most of the point, which in the spring is colored yellow with lupine. A large herd of tule elk lives there, part of the Feds' efforts to restore the ecological workings of the seashore. It seems to me a hopeful task, an attempt to recover some of the area's wildness after 150 years of domestication. I love hiking out there alone on wind-torn afternoons, when the ocean gray is seamless with the fog and the surf rumbles amid the swirls of mist. I like having the chance to watch the elk wrestle with each other. The click-clack of their antlers is bewitching, otherworldly, as if someone were fencing with bone sabers. It's marvelous, that echo from the Pleistocene.

And yet I feel sorry for the elk. Majestic and powerful though they are, they live in a kind of conservation prison. The sea binds them on three sides, and a dairy farm on the fourth. The park service carefully controls their movements, and they can't roam wherever they wish. Occasionally they are rounded up, cattle-like, so that park rangers can manage their breeding to ensure genetic diversity. They aren't exactly wild, and they sure aren't free.

No wonder some critics of wilderness sneer that conservation today amounts to little more than making wild nature into a "museum exhibit." It's depressing to think that we've come to a time in which so few shards of the original Earth remain that we must protect them like we do antiquities. The Tetons and the Sierra Nevada go in the Mountain Collection; the Tsongas and the Olympic belong in the Forest Gallery; the Grand Canyon anchors the Desert Room. But I wouldn't want to live in a world without wilderness any more than I would want to live in a world without museums.

There's a lot to learn from what we might call these "living collections." Perspective, for starters. Just like an ancient vase, a cliff face makes us recognize our provisional place on this planet: many people came before us, and many more will come after. Also, memory. The wilderness vista reminds us where we came from and, in doing so, becomes a vessel of culture, a historical asset. Most important, perhaps, is the intrinsic value of beauty. Only a philistine would ask what a work of art is good for. The exquisiteness of a living landscape doesn't need to explain itself anymore than the Mona Lisa does. Both can leave us speechless, and that's more than enough. (I'm reminded of an old line from Bob Marshall, one of the cofounders of The Wilderness Society. When asked how many wilderness areas the United States needed, he replied, "How many Brahms symphonies do we need?")

The museum metaphor, though, has obvious limits. I don't like living in a world where wildness has to be kept under lock and key. After all, don't we want the wild to also be of this time, of our time?

Maybe it would feel more modern if we imagined wilderness as a time capsule. The National Park Act of 1916, which is about to hit its century mark, declared that the parks were established "to conserve the scenery and the natural and historic objects and the wild life therein . . . and to leave them unimpaired for the enjoyment of future generations." Well, here we are, those future generations. The wildlands that have been preserved were saved for us, for this very moment. The people who saved them were trying to tell us something—something they felt was vital and urgent. So let's enter that time capsule and see what we discover there.

I know this message from the past won't read exactly as how its senders intended. We now know it's impossible to keep any place "unimpaired." Our amazing technologies—universal GPS, anyone?—have filled in the last blank spaces on the map. We can no long take seriously the hope for protecting some untouched Eden. Wilderness was supposed to be protected forever, but we've found that "forever" has a half-life, too.

So then: What would a twenty-first-century wilderness look like?

I don't mean that we should toss out the legal definition of wilderness written five decades ago. It has served us well, and tinkering with it (especially in today's political environment) is only likely to lead to the law's weakening. Nor do I want to compromise the definition of "wild" into meaninglessness by suggesting that the raccoon rummaging through your garbage bin is equivalent to a lynx prowling the deep woods of the Northern Rockies. Such an accommodation of terms just ends up diluting the uniqueness of both the nearby nature and the remote wilderness.

But we do need a new *understanding* of wilderness that can match this new age of the Anthropocene. Our view of wilderness has always been contingent on the times in which we live, and it seems to me it's long past time to update our ideas about the wild. The oyster farm controversy proves as much. We need to find some way to thread the needle between Amy Trainer's romantic yearning for a pristine place and Phyllis Faber's doubts that wilderness even exists.

Can we celebrate wilderness without fetishizing it? Can we put humans back into our mental picture of wilderness and still keep the vision wild? What will it take to craft a wilderness ideal that is at once ironic and heartfelt, a self-aware sentimentality for everything that is wild?

I don't know yet what the new understanding of wildness will look like. But I'm determined to find out.

Let's start at the beginning, at the trailhead.

Yosemite Wilderness

2

The Mountains of California

FOURTH OF JULY WEEKEND, Yosemite National Park, and the traffic was horrific. We were driving the Tioga Road, the serpentine route that leads into the high country, up to Tuolumne Meadows and, beyond that, to the alien-looking shores of Mono Lake.
The road through the granite escarpments looked like some kind of
automotive charm bracelet: fully packed pickups, sedans stuffed with
families, Subarus filled with college kids. There were lots of those
rental RVs that are popular with German tourists. "Cruise America."
the side of the RV says, complete with pictures of the Grand Canyon
and Mount Rushmore.

Our two cars were headed for White Wolf. The plan was to do
a four-day, three-night hike, starting from Tenaya Lake and hiking
down the length of the Grand Canyon of the Tuolumne River. We
would pass the Hetch Hetchy Reservoir and then climb back up

to White Wolf, where we would leave one of the cars at the horse corral.

White Wolf is a high-country summer camp, a mix of drive-in campsites and cabins on the edge of a grassy meadow 8,000 feet above sea level. The cabins are painted in the alpine team colors—plain whitewash with forest-green trim. They rent horses. They have a family-style restaurant with red-and-white-checkered tablecloths that serves far more food than you can eat on a warm summer night. There's a tiny store that sells basic supplies. A sign in front is pitch-perfect. "Kids—we have many flavors of ice cream. Parents—two words: Cold Beer."

That sounded like a good time—watching the wildflowers in the meadow while drinking beer and eating ice cream. But I had other ideas. I was taking my good friends Chris and Alexia on their first trip into the mountain wilderness. The whole thing was a kind of poor man's experiment. I know Chris and Alexia from Alemany Farm, where the three of us have worked together for years, and I was eager to introduce some gardeners to the wilderness. The trip was a pilgrimage of sorts. We would follow old myths through the woods to make new for ourselves the discoveries of others.

There's a whole sub-genre of nature lit celebrating the wilderness's rejuvenating powers. A veteran of the Iraq War finds himself lost upon his return to the United States, rediscovers himself on the mountain summits. A woman suffering from the barrage of chemicals and electronic signals in New York City goes to the woods, sets up camp next to a lake, finds that only there can she think straight. Another woman, a writer, learns that her mother is dying of cancer, seeks refuge among the bird colonies around Utah's Great Salt Lake, and unearths a hard-won equanimity with the universe.

I wanted to collect my own anecdotal evidence—and I wanted the beginner's mind to do so. Ask any longtime wilderness lover what they value about the wild, and often they will struggle for words. The experience defies easy encapsulation, and people will stumble and stutter when trying to explain their feelings. Talking about the wild can be like talking about sex or religion: the power of the thing embarrasses us.

I had planned the hike so that we would retrace some of the same landscape that inspired John Muir's classic book *My First Summer in the Sierra*. What would Chris and Alexia feel during their first excursion in the Sierra? What would they experience when confronted with what Muir called "the vast and glowing countenance of the wilderness in awful, infinite repose"? Perhaps in the first blush of a new experience there would be a spark of romance and my friends would find some kind of magic. Or maybe they would find the whole thing tiresome, uncomfortable, and frightening. I had no idea.

Chris at least looks the part of a rugged adventurer. With his big, black, bushy beard and longish hair swept behind his ears, he wouldn't be out of place in a Gold Rush boomtown. It's a fashionable style these days (think: homeless Don Draper), enough so that Chris had recently been picked off the street by a Levi's scout hoping to capture the zeitgeist. But even if he appears like a prospector, Chris is no outdoorsman. He had never really been in the wild before—though he is, in his way, something of a wild man. As a teenager he had scrapped with the law, then spent a number of summers driving tomato trucks in the San Joaquin Valley to pay for winters spent bumming around Europe. He had crisscrossed the United States by motorcycle, and in the course of his travels he had visited plenty of national parks, where he camped next to his bike. Yet he had never been more than ten miles beyond blacktop in his life. In an e-mail he sent me before we left, he said he was "excited and nervous."

His wife, Alexia, was a more experienced hiker. Her stepfather had been a colonel in the British Army, a connoisseur of the lowland heath (walking sticks, old stone walls), and growing up she spent many days ambling the trails of Devon. On one trek, when Alexia was a little girl, he made her tiptoe out onto the soft sponge of a bog to retrieve the perfectly preserved skull of a deer that had got stuck and died there. If that sounds unnecessarily dangerous, know these two things: One, he *did* tie a rope around her. And, two, Alexia is an uncommonly small woman. She would later run off and join the circus in Spain, where she worked as a trapeze artist, and where, in Barcelona, she met Chris. As an adult, she can't break 100 pounds. As a child, she must have weighed nothing. There's no way she would have gotten stuck in the moor.

Also in our crew was my friend Michelle. She's a wiry, athletic woman, a nutritionist at her day job and a hiker, jogger, cyclist, and yogi the rest of the time. Since Chris and Alexia were newbies, I was glad to have another experienced backpacker on the trip. In college she had worked as a wilderness guide, and I knew she could fit a pack on someone better than the staff at REI.

After doing a final gear-check at White Wolf, we left my car there and piled into Chris and Alexia's station wagon for the twenty-minute drive to Tenaya Lake. The scene at the long lake was a snapshot of summertime glory. A large Latino family was chilling at the lakeside, the kids splashing in the cold water as the parents and grandparents hung back in beach chairs in the shade. Two hipsters were rigging a Sunfish, one of those little sailboats made for freshwater lakes; for some reason, one guy was wearing only purple briefs. A group of Thai tourists were taking pictures in front of the lake, the sharp white point of Tresidder Peak framing the background. One woman was wearing heels and a black cocktail dress. Out of the corner of my eye I spied a team of climbers a few hundred feet above us, tackling the smooth granite face of Polly Dome. A single climber and his rope line were silhouetted between the rock and the sky, forming a kind of adventurer's exclamation point.

The four of us took in the scene over a quick lunch of pita, sardines, cheese, and some plums that I knew wouldn't survive the crush of my pack.

I shared what I knew about the lake:

When the whites cleared the Ahwahnechee Indians out of Yosemite (which, incidentally, means "those who kill" in the language of the Ahwahnechees' ancient enemies, the Miwok), they told the chief, Tenaya, that they would name the lake after him. He said, "The lake already *has* a name. We call it Py-we-ack." The whites called the lake "Tenaya" anyway.

Finally ready, we decided to hit the trailhead, located just across the road. We climbed above the traffic and some twenty yards later were greeted by a carved wooden post: "Yosemite Wilderness."

Beyond that stood a more officious metal sign, "Entering Yosemite Wilderness," and a list of rules and regulations, among them, "Help to protect the wild animals by keeping them wild."

I have always loved these signposts located at the edge of federally protected wilderness areas. They are so storybook in their certainty—"Entering Wilderness"—as if there were some bright line demarcating the wild and the tame. It's like the map from *The Hobbit*: "The Wild →." Cross this border, the sign at once promises and warns, and things are guaranteed to be unpredictable.

We started humping up a dry streambed called Murphy's Creek, and it wasn't long before I was feeling the effort. The grade upward was steep, and the trail, though well defined, was rocky and rough. The pack pushed into my hips, and the shoulder straps creaked with every step. Flies and mosquitos flitted about my sweaty face. I could feel the altitude, the thinness of oxygen at 8,300 feet.

The trail climbed for a couple of miles and then, as we crossed a small saddle, evened out. We cut across a sweep of open granite, the trail marked by cairns, and threaded our way between two small peaks. We slipped through a stand of pine and then—*bam!*—the landscape blew wide open before us. As far as we could see, nothing but forest and the spires, peaks, and domes of rock: Ragged Peak, Sheep Peak, Excelsior Mountain at the edge of sight. "Mountains beyond mountains," Muir had written, "glorious forests . . . sweeping over countless ridges and hills, girdling domes and subordinate mountains . . . filling every hollow."

Chris was just as impressed, if less poetic: "Man, this country just goes on forever. It feels like there's no one around. Like we're the only people out here." I didn't have the heart to tell him that the Tioga Road continued somewhere through the scene; from our vantage point the asphalt was cloaked by trees. Civilization might have slipped from view, but it wasn't all that far away.

We were just about to start on again when Alexia looked Chris in the eye and said flatly, "I'm going to be sick."

Then she hurled.

We were stunned. Michelle and I looked at each other wide-eyed. I looked at Chris to get a clue about what was happening. Alexia hurled again, vomit blasting onto a bed of pine needles. Remember that scene in *The Exorcist*? It was kind of like that.

"Oh my God," Michelle said. "She's got altitude sickness."

Exactly. There was nothing wrong with Alexia. She hadn't eaten anything that the rest of us hadn't eaten as well. She hadn't complained of a fever or nausea. It was simply that her body wasn't getting enough oxygen—and now her body was in revolt. At just above 8,000 feet, we weren't all that high, but the elevation, combined with the exertion from the climb, had been enough to wreck her. She stood leaning on tree, looking weak-kneed and frightened.

There is really only one thing you can do about altitude sickness: drop elevation. And that's what we were planning on doing anyway. This lookout was at the highest point of our trip. During the next three and a half days we would drop about 4,000 feet as we followed the course of the Tuolumne River downstream. Our best course of action was to keep going forward.

Michelle and I told Alexia all of this. We told her that she should drink plenty of water and take it slow and steady. It was still the early afternoon and we had hours of daylight left. We only had to make it another four miles, just past Tuolumne Falls, and then, according to the conditions of our wilderness permit, we could camp anywhere. "We're going to go real slow, Lex," I said. "We've got all the time in the world."

She nodded quietly and sipped some water. I looked over at Chris. "She's okay," he said, though I'm not sure either of us believed it.

I made sure to set gentle pace, and we took it slow as we made our way down the well-worn path. At our leisurely rhythm it was easy to pick up on the nuances of the landscape. As we came into the small depression of Cathedral Creek, the woods took on an extra lushness. The large, white, umbrella-like heads of cow parsnip and ranger's button crowded among the grasses. Purple whorls of large-leaved lupine brightened the scene. Alexia perked up and found the energy to take some pictures beneath a massive western juniper, its trunk about fifteen feet round, its cinnamon-colored bark glowing red in the afternoon light.

Large, dark thunderheads had been shadowing us for miles, every once in a while letting out a growl of thunder. Now we could hear a roar coming from ahead of us: the sound of Tuolumne Falls. We came down to the falls and stopped to watch the water tumbling and tumbling over a rock ledge. I was mesmerized, as I always am, at the contradiction of a cataract: an always-moving flow whose shape is ever-constant. A thing at once speeding and still.

On the far side of the pool we could see some of the white tents of the Glen Aulin high camp—sort of like White Wolf, only you have to hike about six miles to get there. Many of the campers were at the pool's edge, some reading, others just sitting hypnotized by the water.

We moved on and, after crossing a pair of foot bridges, came to the top of a ridge line. The Grand Canyon of the Tuolumne spread before us, a vast V of rock split by the river. Ridge after ridge stretched to the western horizon, the cliffs rising a thousand feet above the river. The curve and sweep of the granite looked like waves, the solid version of the waterfall behind us. Down in the valley, aspens gilded the ribbon of the water, and some of the ponderosa looked to be well over 100 feet tall. Chris made a pair of devil's horns with his hand and stuck out his tongue. In case you didn't grow up in Orange County, that roughly translates as "fucking awesome."

We started to climb down the switchbacks in the rock face. The trail turned sandy along the river's edge and I started to groove along, feeling good as a soft breeze came up the valley. Then I noticed no one was behind me. I stopped and waited, and waited some more. I was just about to start back when I saw Alexia hurrying along the trail. "Michelle sprained her ankle," she said breathlessly as she caught up to me.

I hurried back along the trail until I came to where Michelle was sitting on the ground, massaging her left ankle. It turned out that she had been walking oddly all afternoon, trying to get used to a new pair of boots she had bought the week before. She had been favoring one foot until she twisted the other. When she stood up, she was limping. I grabbed Michelle's pack, offered her my arm, and helped her down the slope.

Clearly it was time to call it quits for the day. Chris and I jogged down the trail to scout for a campsite, and about a quarter of a mile from the bottom of the ridge we found a nice spot under an ancient juniper—flat ground and a well-used fire ring. I went back to collect Michelle and her gear. Soon she was perched on the riverbank, her swelling foot resting in the cold water.

The four of us spent the rest of the afternoon at the river's edge. As the sun made its final descent, some of the Glen Aulin campers gathered at the top of the valley to admire the sunset. We watched the sightseers watching the scenery, the dying light turning the canyon walls into swirls of rose and red. As dusk turned to dark we stayed on the riverbank, just taking in the world. Swallows came out of the shadows and zigzagged over the water hunting for insects, as sharp and as quick as fighter jets. It seemed that maybe everything would be all right.

But dinner was another miniature misadventure. As soon as she caught the first scent of food, Alexia puked again. She managed to hold down a bit of plain rice, then hurried to bed. Michelle wasn't far behind, and as she made for her sleeping bag she was tiptoeing gingerly. Chris turned in soon after. I was left alone by the small campfire, silently watching the logs turn to embers.

I felt bad for my friends, but I'll admit I was annoyed and disappointed. It looked like this little excursion into the wild would be over as soon as it had begun.

The original American ideal of wilderness that we had gone in search of was forged on the mountaintops of California's Sierra Nevada. John Muir called the place "The Range of Light," and the vistas of sky and rock in the high alpine reaches inspired him to impassioned poetics that forever changed the way many people thought of the wild. In the late nineteenth and early twentieth century Muir became the foremost prophet of wilderness, and a century later he remains the epitome of the wildman-ecstatic.

Muir was an odd bird. The son of Scottish immigrants who had settled in Wisconsin, he was a freethinker despite (or because of) a harsh upbringing by a religiously fanatical father. At the age of twenty-six he skipped across the Canadian border to dodge the Civil War draft, and three years later he walked 1,000 miles from Indiana to Florida, collecting botanical specimens along the way. He was an ingenious mechanic with several patents to his name and an amateur geologist who fundamentally altered the scientific understanding of how mountains and valleys are formed. Above all, Muir was a self-fashioned mystic who discovered that amid wild places he could know God better than he could from reading his father's Bible. "The clearest way into the Universe," Muir wrote, "is through a forest wilderness."

Of course, Muir wasn't the first person to find God in nature. When Muir arrived at Yosemite he carried with him a tattered copy of Ralph Waldo Emerson's *Essays*. Emerson was the main intellectual force behind American Transcendentalism, a philosophy based on the idea that there is a plane of reality, and a realm of spiritual truth, higher than the physical world. The best way to glimpse this transcendental space was through nature—or Nature, as Emerson would have written it. "In the wilderness I find something more dear and connate than in the streets or villages," Emerson had written.

Emerson's most famous disciple was Thoreau. The small cabin Thoreau built on Walden Pond was supposed to be an oasis of wilderness "in the desert of our civilization." For Thoreau, wilderness was a spiritual and intellectual tonic: "The poet must, from time to time, travel the logger's path and the Indian's trail, to drink at some new and more bracing fountain of the Muses, in the far recesses of the wilderness."

Pay special attention to one key phrase there: *from time to time*. On close inspection, it becomes apparent that Thoreau's views on nature and civilization were more complex than the caricature of the wilderness hermit we hold in our minds. His cabin was just a short walk from the town of Concord and within earshot of a railroad line. During the time he lived at Walden Pond, he was a frequent guest at Mrs. Emerson's table. And when Thoreau did go beyond the rather tame woodlands of Massachusetts, the experience wasn't always as transcendentally wonderful as you might expect.

Listen to Thoreau's impressions upon climbing the summit of Maine's Mount Katahdin. "A place of heathenism and superstitious rites—to be inhabited by men nearer of kin to the rocks and the wild animals than we," he wrote. The rawness of the landscape unhinged him. "What is this Titan that has possession of me? *Who* are we? *Where* are we?"

Thoreau liked his wilderness in mild doses; he was more a fan of the pastoral ramble than of trailblazing. Eventually he concluded that humanity's relationship to nature should be a kind of middle path: permanent residence in a "partially cultivated country," with occasional excursions into the city and the wilderness as touchstones for art and spirit. "Nature has a place for the wild clematis as well as for the cabbage," he wrote in his essay "Walking."

It would take Muir to make the unbridled case for the divinity of the wild. When he disembarked at the San Francisco waterfront in 1868, someone asked him where he was headed. "Anyplace that is wild," he answered. Then he walked several hundred miles across the San Joaquin Valley to the cliffs and canyons of Yosemite, where he discovered a wilderness that appeared to him the evidence and expression of "a good God."

Muir's mountain rhapsodies punctuated a century in which the popular attitude toward wilderness went from fear and loathing to praise and wonder. A generation before Muir was born, British Romantic poets such as Lord Byron, John Keats, and William Wordsworth began to celebrate the virtues of wild nature. Here's Keats, glorifying the splendors of nature in his poem, "To One Who Has Been Long in City Pent":

'Tis very sweet to look into the fair

And open face of heaven,—to breathe a prayer

Full in the smile of the blue firmament.

Who is more happy, when, with heart's content,

Fatigued he sinks into some pleasant lair

Of wavy grass. . . .

Given the ancient fear of the wild, the Romantics' appreciation for wilderness marked a major reversal in Western thinking. "With the flowering of Romanticism . . . wild country lost much of its repulsiveness," writes Roderick Frazier Nash in his authoritative book, *Wilderness and the American Mind*. "It was not that the wilderness was any less solitary, mysterious, and chaotic, but rather in the new intellectual context these qualities were coveted." It took the vices of the industrial city—"dark Satanic mills" and soot-covered streetscapes—to make the wild into a virtue. While the city was gross and wretched, the untamed vista was *sublime*, to use one of the Romantics' favorite words. Here's how an early-nineteenth-century American surveyor defined the term: "The Sublime in Nature captivates while it awes, and charms while it elevates and expands the soul."

The romanticization of wilderness quickly jumped the Atlantic. Early American citizens were beset by a cultural inferiority complex and hungry for local distinctions to match the accomplishments of Europe. In the awesome natural features of North America they found something to brag about. Eager to create a new history for themselves, white Americans embraced the wild as an Eden out of time.

In the nineteenth century, the wilderness ideal was most often expressed on canvas. Painter Thomas Cole created wilderness scenes that thrummed with drama: dark storm clouds shadowing a vacant forest, Niagara Falls touched with a heavenly light, as if the wild were being baptized. Cole influenced later painters like Frederic Church and Albert Bierstadt and helped to create a uniquely American style of painting. These painters worked in a single key—magnificence. In their paintings the entire world was grandeur.

Muir then elevated wilderness appreciation to the next level. He took the Romantics' poetry, stirred in the painters' bombast and the mysticism of the Transcendentalists, and combined it with scientific inquisitiveness and a crypto-Christian fervor that resonated with millions of his fellow citizens. Muir was a major author of his day, and books such as *The Mountains of California*, *My First Summer in the Sierra*, and *Travels to Alaska* were best sellers. "Climb the mountains and get their good tidings," Muir told his readers. "Nature's peace will flow into you as the sunshine into the trees."

By the standards of our ironic age, Muir's ecstasies can be a bit too rich and frothy. But remember that this was someone who as a child was forced to memorize the entirety of the Old and New Testaments. The rhythms of the King James Bible were seared into his mind, making him a perfect messenger for connecting with a God-fearing people. "Every hidden cell is throbbing with music and life," Muir wrote of Yosemite Valley, "every fiber thrilling like harp strings, while incense is ever flowing from the balsam bells and leaves." With his long beard and walking stick, Muir was like a latter-day John the Baptist, bearing salvation out of the wilds. Muir made it safe for a Christian nation to embrace the wilderness.

Even in the midst of his reveries, Muir was blazing ahead with scientific discoveries. The extreme conditions of the Sierra Nevada offered a perfect laboratory for examining the environment. A grove of eighty-foot-tall pines thriving in a slim granite notch demonstrated how soil forms. The vast valleys revealed the eons-long work of lava and glaciers. The interplay of animal, vegetable, and mineral led Muir to perceive the interconnection of all things, which is the central insight of ecology. Walking through the woods around Tenaya Lake, Muir thought, "When we try to pick out anything by itself, we find it hitched to everything else in the universe."

I've trekked many miles through the Sierra Nevada—Kings Canyon Wilderness, Desolation Wilderness, Carson-Iceberg Wilderness, Emigrant Wilderness, Yosemite Wilderness—and the mountains have enthralled me, too. Peaks are formed by fire, then worn smooth by water and ice. Epochs engrave cracks into the cliff faces and the seasons draw lines in the rock. Then, with each hour, with every shift of sunlight or moonglow, the scene is remade into something unprecedented. Few other landscapes inspire such an intense simultaneous experience of ancient past and visceral present. To climb to the top of the world, where there is nothing between you and sky, can feel like brushing celestial grace itself.

As I sat by the dying fire at our campsite at the head of the Grand Canyon of the Tuolumne, I hoped my friends would get the opportunity to catch a glimpse of the sublime. Between the sprained ankle and the altitude sickness, that didn't seem likely.

As soon as I woke up I could feel the closeness of civilization. We weren't more than a mile from the camp at Glen Aulin. If we needed help with Michelle's ankle, assistance was a short hike away. But going back upriver was the last thing I wanted to do. Every bit of me wanted to keep hiking downriver, further into the wilderness. The mere thought of Glen Aulin was claustrophobic, like being cornered by a close-talker. I knew, though, that I couldn't just listen to my instincts. We would do whatever was needed to make sure everyone was okay.

Happily, Alexia was feeling better. A good night's sleep acclimated her to the altitude, and before I was out of my tent she and Chris were having a breakfast of oatmeal and tea. The real problem was Michelle's ankle. She tried putting her boot on, but it was no good—she couldn't walk with it on. We seemed to have only a few options: walk up to Glen Aulin and see if there was a park ranger who could help; hike back upriver, through Tuolumne Meadows to the Tioga Road, a trek of about five miles, and hitchhike back to the cars; or spend a few days at our camp just hanging out. Continuing downriver didn't seem a choice. We still had about twenty-five miles to go, and going onward would only take us away from any assistance.

But Michelle said she wanted to keep going. "Staying here for three days sounds boring," she said. "Let's do this thing."

So we came up with a work-around. We made a stirrup with some medical tape, crisscrossing the tape underneath the arch of her foot and around her ankle. Over the medical tape we bound her foot and ankle in a second stirrup of ace bandage. In place of a boot she would wear a Tom, one of those canvas flats that resemble the sandals people in Nepal wear. She looked weird—a hiking boot on one foot, a canvas sandal on the other—but it worked. Bolstered by a solid walking stick, she could hike.

The sun was angling toward noon as we set off down the river. The trail was flat, sandy, and easy to walk as we headed through the canyon. Soon enough my worries drifted away as I slipped into the rhythm of the world around me.

The canyon rose above us, rock walls as tall as a skyscraper. The gray granite face was threaded with green, creases in the rock where juniper and pine had managed to find a root-hold. In the river bottom, summer was in full swing. Wildflowers were everywhere, the meadows daubed with the yellow of Sierra butterweed and the orange of the alpine lily. Purple was the most popular shade: purple lupine, the delicate white and purple cups of mariposa lily, purple mountain asters, and clutches of purple harvest brodiae, long-stemmed and six-petaled. Aspens edged the riverbanks, and when the wind stirred the leaf-rustle was so strong that I mistook the sound for the rush of a waterfall.

Eventually we came upon the falls. First was California Falls. The river there gathers into a long pool, placid and pine-shaded, then spills over a rib in the rock to become a white plume, the water descending like a staircase. A mile or so later we hit Le Conte Falls. There the cascades are broad and long, the river sluicing down a stone slope like a water-park slide. Finally, another mile farther along, Waterwheel Falls. The scene there is similar to Le Conte—the river rushing over a long granite slope—only the force is greater, so that the water at times curls up and over itself, seeming to spin in place. The waterwheels rumble for close to half a mile, the sound so big that we had to shout to make ourselves heard.

We continued on, the river always on our left, and as the canyon leveled out we came into a grove of burnt sugar pines. The sugar pine is the tallest pine in the world, sometimes growing to more than 200 feet, and the trunks were massive—huge blackened pillars stretching upward. The fire looked as if had occurred just the summer before, and everything was dead except the trees. The ground was dark with soot, the only hint of color the unbelievably huge, caramel-colored pinecones scattered in the black dust. A hush seemed to fall on the forest, a sepulchered silence. "It's so *Game of Thrones*," Chris said, and we all chuckled.

The reference didn't surprise me. We have become such strangers to wild nature that we don't know how to process it, and often when people enter the wilderness for the first time their imaginations fall back on fantasy. The mind scrambles for some kind of analogy, even if that means resorting to fiction and fables. *City Slickers* is a reference I've heard more than once from a first-time backpacker, ditto *The Princess Bride*. I may have more backcountry experience than most, but I'm not immune to this tic of the mind. I've spent my share of trail miles walking through Tolkein's Middle Earth, my experience filtered through the memory of invented landscapes. In the modern world, it seems, most of us only know the real by way of make-believe.

The sun began to dip. Michelle was starting to lag. She had been a trouper all day, enduring the rocky descents past the falls and putting up with the comments of hikers we passed on the trail ("New technique?" one woman said drily when she saw Michelle's footwear). Now she was hitting her limit. I began to keep an eye out for a good place to camp.

I was in the lead, walking fifty yards in front of the others, when the trail turned to reveal a scene that was perfect in its accidental symmetry. Through some trick of geology the vast canyon had turned into a crib of rock: the granite walls rose up to the north and south, and directly to the west there was a third cliff face, the river bending invisibly beneath it, so that I felt surrounded by stone. A no-name waterfall on the south side of the canyon completed the arrangement. The cascade plummeted down the cliffs, first one veil and then another, columns of water sliding through the air.

Then I felt it—the tingle of the sublime. The feeling of "dumb admiration," Muir called it, when we experience an "infinite mystery." The quiet of the canyon went even quieter as a silence stole out of the ether. I couldn't tell you what it was, or even if it was real. Maybe it was a trick of light. Maybe it was just the subtle vibrational change of the stomata on a thousand leaves and needles opening or closing with a shift in barometric pressure. Maybe it was a delusion conjured by desire. But I know that I felt enlarged, like a string had loosened in my heart, a knot that I didn't even know was there.

Footsteps broke my reverie: Chris and Alexia and Michelle coming up the trail. When they caught up to me they stopped and took in the scene—the canyon's sudden coziness, the waterfall pouring down the rock—and we decided this was the place. Not far off the trail we found a sandy ledge with a fire ring set amid manzanita and the bright yellow blossoms of sulfur flower and hoary buckwheat. There was a small beach at the river's edge. "This place is epic," Michelle said, and we dropped our packs for the day.

Michelle's pluck while hiking the Grand Canyon of the Tuolumne made me think of the second major strain of American wilderness appreciation—that of the explorer and the adventurer. If some people enter the wilderness in search of the transcendent, others go for the challenge of a place that feels more alive than the domesticated landscapes of city and farm.

When Muir was at the height of his popularity, Americans were experiencing an identity crisis. In 1890, the US Census Bureau announced that the frontier was officially closed, news that came as a blow to the American psyche. A prominent historian of the time, Frederick Jackson Turner, warned that the disappearance of the frontier wilderness threatened the loss of much that was best in the American identity: self-sufficiency, self-confidence, and the spirit of self-governance. The pioneer's struggle in the wilderness had been key to creating a democratic culture. "The very fact of the wilderness appealed to men as a fair, blank page on which to write a new chapter in the story of man's struggle for a higher type of society," Turner wrote in a widely discussed essay published in the *Atlantic Monthly* in 1896.

Contemporary historians have criticized Turner for overstating his case. The frontier myth was just that—a story we tell ourselves to explain our identity. The pioneer experience had always been about cooperation as much as rugged individualism, the westward expansion as reliant on communal *inter*dependence as *in*dependence. Also, the continent was never a "blank page." Other people had lived here before the arrival of Europeans. If North America appeared empty, that's because the original inhabitants had suc-cumbed to disease or were driven from their lands.

Still, a myth doesn't have to be true to be powerful. Millions of Americans believed (and still believe) that the frontier wilderness was the crucible of the country's character. The prospect of losing the wilderness sent shockwaves through society. A "sense of nostal-gic regret over the disappearance of wilderness" took hold of the country, Nash writes. Until the crescendo of industrialization and urbanization of the late nineteenth century, many Americans lived with wilderness just outside their front doors. They thought of the wilderness as an obstacle to overcome. As the wilderness was steadily conquered, though, popular opinion changed.

Hikers and mountain climbers formed organizations to protect their outdoor playgrounds—there was Muir's Sierra Club in the West and the Appalachian Mountain Club in the East, as well as the Campfire Club of America. In 1910, the Boy Scouts of America was established. Its *Handbook* bemoaned the "growth of immense cit-ies" and declared that happiness was most common among "those . . . who live nearest to the ground . . . who live the simple lives of primitive times." In the first thirty years of its publication, the Scouts' *Handbook* was the second-best-selling book in the United States, topped only by the Bible.

Conservation was becoming a political force. In 1872, President Ulysses S. Grant signed into law a bill creating the first government-protected wilderness preserve in the world—the 2-million-acre Yel-lowstone National Park in Wyoming. In 1885, the New York Assem-bly established a 715,000-acre "Forest Preserve" in the Adirondack Mountains, later expanded into a 3-million-acre state park (today it's twice that size). These first preserves were, in a way, acciden-tal achievements. Yellowstone's preservation was largely driven by railroad interests looking to cash in on tourist traffic to the area's unique geysers. The Adirondacks were conserved in order to safe-guard New York City's water supply.

But the new parks weren't just protected for the sake of narrow self-interest. The Transcendentalists' ideas infused the political discus-sion over wilderness conservation. In an 1886 Congressional debate over allowing a railroad to pass through Yellowstone, a congressman from New Jersey defended Yellowstone as one of those "mysteries of nature that elevate mankind and bring it closer to communion with omniscience." (The railroads lost the vote, by the way.) During an 1894 convention to rewrite New York's state constitution, one speaker defended Adirondack State Park saying, "When tired of the trials, tribulations, and annoyances of business and everyday life in the man-made towns, [the Adirondacks] offer to man a place of retirement."

No national figure embodied the era's lust for wilderness like Theodore Roosevelt. Our twenty-sixth president, born into a wealthy family in New York City, was a sickly child, afflicted by asthma and other bronchial ailments. After attending Harvard and serving in the New York Assembly, he went west and became a rancher in the Dakota Badlands. There he reinvented himself. He bulked up, dressed in buckskin, spent much of his time hunting and riding. He helped shape, and then became a devotee of, Frederick Turner's thesis that the frontier was the foundation of America's democratic culture. "Under the hard conditions of life in the wilderness," Roosevelt wrote in his 1889 book, *The Winning of the West*, those who came to North America became new men "in dress, in customs, and in mode of life."

While Muir married the wild to the divine, Roosevelt hitched wilderness to Americans' idea of themselves. During his presidency, Roosevelt urged his fellow citizens to embrace a "life of strenuous endeavor" and declared that national parks were "essential democracy" at work. On a cross-country tour in 1903, he visited Yellowstone and Yosemite, where he spent several nights talking around the campfire with Muir. One morning they awoke to find themselves dusted with fresh snowfall, an event that Roosevelt called "one of the most pleasant nights of my life."

Some of TR's rhetoric can be hard to swallow nowadays. As the privileged scion of an old Manhattan Dutch family, Roosevelt was blind to the fact that most Americans were unconcerned about succumbing to "slothful ease"; the country's farmers, miners, and cowboys were working plenty hard in the outdoors, thank you very much. His celebration of wilderness as a font of "manliness" sounds especially anachronistic today.

But some part of Roosevelt's "Bully!" spirit remains alive and well. You can see it among the adrenaline junkies who tackle the wilderness as a physical challenge—the rock climbers, the back-country skiers, the kayakers, and trophy hunters. All of them are, like TR, living a life that has a "scorn for discomfort and danger." They go into the wilderness to test themselves—to toughen the body and, in the process, strengthen the spirit.

Of course, one doesn't have to be a macho dude-bro to pursue such adventures. You don't even have to be a dude: the most popular wilderness book of the last decade was written by a woman. During her misadventures along the Pacific Crest Trail, Cheryl Strayed found the experience of day-after-day hiking to be as tough as her divorce and the death of her mother. "The experience was powerful and fundamental," she writes in her memoir, *Wild*, as she expresses surprise at "how profoundly the trail would both shatter and shelter me."

I've felt that power myself, and for me the wilderness also provided an opportunity for self reinvention. Like Roosevelt, I was a weak kid. I suffered from asthma, hay fever, and allergies. I wasn't athletic, and though I won the "Most Inspirational" trophy on my junior high flag football team, that was mostly because I was the smallest player with the biggest voice. Nor was I physically courageous. Like all the other kids in my neighborhood, I was a skate-boarder. But whenever it came time to hit the half-pipe, I concocted some lame excuse to bail: "Um, my mom needs me to, uh, clean out the carport."

In the wilderness I found strength and courage. I discovered that I could walk far and fast, and that I could do so with a heavy pack and hardly notice. I wasn't afraid of the woods or whatever lived there, nor was I afraid of the dark or of getting dirty. I didn't mind slogging up a muddy trail or walking miles in the rain. I liked it, in fact. I liked feeling myself rubbing up close to the earth.

I couldn't tell you what exactly drew me to the wilderness. When I was growing up in Arizona we never went backpacking, and the few times our family did go car camping it usually ended in some kind of snafu. The wilderness didn't play a role in the landscape of my childhood.

As I got older, though, I began to feel an inexorable pull toward the untamed outdoors. So, in my sophomore year of college, I asked my parents for a backpack. Spring break was approaching, and some friends and I had made plans to spend the week hiking part of the Appalachian Trail in Virginia. None of us had ever done anything like that before, but we were young and bold enough not to care.

A few days before we were going to leave, I took the time to organize my gear and see what the pack felt like on my back. I was walking up and down the hallways of my dorm with the pack strapped on when I ran into the floor RA. He was one of those guys that my friends and I called a "white hat"—our shorthand for the rich kids who had gone to prep school. He was a jock, a member of the water polo team. A few months earlier he had busted me and my pals for smoking pot in the dorm. We didn't really like each other.

He asked me what I was up to, and when I told him that I was headed out for a week on the Appalachian Trail, he flinched with surprise. "Wow," he said, obviously impressed. It seemed the idea was outlandish to him. Though strong and fit, he wasn't the kind of guy to go into the woods. Then I watched him look me up and down, and as I stood there with my backpack, I very clearly saw him do a full reappraisal: Maybe this kid isn't just some skinny stoner. "Cool," he said.

I headed back to my room with a smile. Even before I had spent a day on the trail, it seemed the wilderness had made me into a new man.

On Day Three of our trek down the Tuolumne River the water filter broke. My water pump had lasted a good ten years and had traveled with me to wildernesses across North America. Now it was done. A small stress fracture had appeared on the piston. Nothing more than a hairline crack, but it was enough so that water sprayed out of the side with every pump.

This was a problem. It's best to treat the water in the backcountry, the biggest risk being giardia, a parasite transmitted through animal feces. We could boil all of our water, but that would be hugely time-consuming and would deplete our fuel supply. We could go ahead and drink the river water, but that would be risky. Michelle had suffered giardia before, and she was adamant that she wouldn't drink untreated water.

We had to find a way to fix the purifier. Sitting on the pebble beach next to the river, Chris disassembled the thing. He scrubbed the filter so that it would pump as easily as possible. He lubed it. Then he put it back together and used some electrical tape I had in my pack to patch the fracture. It was still a mess—water gurgled through Chris's hands with every pump—but it would work.

My bigger worry was Michelle's foot. She could walk as long as the tape and ace bandage stirrup held up, but without the extra support she limped. Unfortunately, we had used all the medical tape we had. The night before, we had carefully saved what was left, spooling it around a water bottle. Now, as we rewrapped Michelle's foot, it was clear that the elasticity and adhesive were wearing thin. Resetting the stirrup, Michelle and I looked each other in the eye, our silent concern obvious. We still had about fourteen miles to go.

As we descended farther into the canyon, the river still on our left, the environment shifted. The landscape was becoming drier. Manzanita was more common, and whenever the river cut through a gorge and the trail climbed up the side of the canyon, we found ourselves amid great hedges of yerba santa, the gray-green, lance-like leaves giving off a cloying scent. New flowers showed up—the purple caps of mountain pennyroyal, the magenta down of pussy-paws keeping close to the ground, as if to escape the sun. Oaks began to appear: blue oaks on the arid slopes, and black oaks and valley oaks in the flats near the water. It was much hotter than the day before. Soon we had all worked up a good sweat.

Sometime after noon we came to a lovely little waterfall pouring off the north side of the canyon. A narrow pool had formed beneath the cascade, a deep reservoir of incandescent green. It was a natural oasis, and the spot was busy with other backpackers. We hadn't seen that many people on the path since Glen Aulin, but, as sometimes happens in the backcountry, a kind of a trail society had formed among us and a few other groups of hikers as we passed them and they passed us, our routes woven together by our walking rhythms.

We had christened one group "Stanford"—three thirty-some-thing women, all very intense-looking, athletic, and expensively equipped. As we sat on the rocks at the edge of the water, our skin drying after a dip in the pool, I talked Michelle into approaching the "Stanford" women to ask them if they might have any athletic tape. One woman rummaged in her pack, pulled out a large roll of tape, and said we could have the whole thing. The tape, so clean and pure white, seemed like a minor miracle. A little ball of worry that I had been carrying in my gut for the last two days melted away. Michelle would be fine.

We reluctantly put our clothes on and heaved our packs onto our shoulders, Michelle now in the lead. We hadn't hiked far when the rock walls began to spread apart and shrink in size. As the canyon opened, the river slowed. The water no longer thundered, but just kept up a steady burble alongside us. We were coming to the end of the Grand Canyon of the Tuolumne. As we entered a place called Pate Valley, the terrain flattened and suddenly everything was very lush. Tall stands of thick, green grasses covered the forest floor. The air became muggy and buggy, and the trail turned muddy in the lower spots. Glades of bracken closed around the path, the yellow-green fans of the ferns swishing against our packs as we passed through them.

The sun was in the west, and it was time to think about finding a campsite. Michelle was doing fine, but now Chris looked in bad shape. He was having a serious blister issue, and I felt bad watching him hobble. Finally I glimpsed an opening in the pines—about twenty yards off the trail, right next to the river, a nice, flat spot, open to the sky and with a fire ring.

That night I decided to ask the new backpackers about their first wilderness adventure. I had avoided talking about the wilderness ideal because I didn't want to skew Chris and Alexia's experience with overthinking. I wanted them to keep their beginner's mind, to really *feel* the wilderness without being encumbered by too many ideas or too much history. I wanted their wilderness experience to be—I guess you could say—*pristine*.

The wilderness ideal isn't obvious or intuitive. Nor can an appreciation of wildness be easily taught—it has to be earned amid the trials of the trail. Like Thomas Paine's idea of revolution, an appreciation for wilderness has to be renewed and refreshed by each generation, which will, inevitably, understand the wild in a new way. The Romantics had rescued wilderness from the Puritans, and then the post-pioneers of Roosevelt's age put it on a pedestal. Now, here in the Anthropocene, can we rescue wilderness once more, save it from a blinkered romanticism and keep the ideal meaningful?

A campfire has a way of loosening the tongue, the primal dance of the flames acting as a social lubricant. A flask of bourbon helps, too. As we sat around the fire after dinner, I asked Chris and Alexia what they thought about the past few days. One by one they ticked off the same feelings that had inspired Muir and Roosevelt a century before, only with twenty-first-century twists.

"Most people will never see this, because it's not easy, and they want it to be easy," Chris said as the flask went around the fire. "Being out here, it challenged me to behave differently. I had to be more patient. It was like—this is happening, deal with it. Lex's altitude sickness. Michelle's ankle. The water pump breaking. You just have to be flexible, or you won't survive. You have to learn how to be out of control. You have to accept it. That's just how it is." He chopped the air with his hand: "You. Are. Not. In. Con-trol."

I threw a log on the fire. The conversation turned to the way that the wild returns experience to life's basic elements. For Chris and Alexia, the stripping away of the unnecessary had been huge—having to survive in the world unassisted by the conveniences to which we've become accustomed. This is, of course, one of the classic attributes of the modern wild, and for Chris and Alexia it had felt revelatory. "All you have to do out here is take care of yourself," Chris said. "At home you barely have to do that. You barely have to think about it. And when it does become complicated—like cooking—people are like, 'Forget about it.' Out here, it's just simple."

"There are no shortcuts here," Alexia said. "It says in the park brochure that it challenges you. It does. It changes you. It opens your mind. It's mind blowing. I never would have thought before today that I could have done this. There's something empowering about it."

Everyone was quiet. A log popped in the fire, sending sparks up into the night. Chris said, "It's not about what I'm trying to get away *from*, but what I want to get away *to*. It's liberating here. At home, everything is so controlled. It's like, 'you can't do that.' And I'm like, *arghhh*. Out here, no one cares. I can jump naked in the water or get dirty and it doesn't matter. Because I'm free."

That probably would've been enough for old TR. Then Chris took it a step further, and talked about the unique Americanness of the wilderness experience. "It also made me kind of patriotic," Chris said.

Most Americans have gotten over the old inferiority complex regarding Europe by now. But for Chris it still rankles. Alexia is a true English Lady. That is, she's descended from aristocracy: she's the great-great-great-great granddaughter of the Duke of Wellington. I think it's fair to say that her family couldn't quiet figure it out when she married an American truck driver. For Chris, the wilderness was one way in which the United States has bested old England. He said, "I was thinking, I am so glad that I can share this with Lex, since they don't have this in Europe. This country has done some good things, and this is one of them."

Then the moment I had secretly been hoping for occurred—a confessional of the sublime. Alexia was the one who said it: "I was sitting down by the river this evening, and my eyes started welling. It was just so beautiful. I was crying because it was so beautiful and because I was happy to be here."

She stopped, smiled, and, it seemed, remembered that she was English. "I surprised myself, because I didn't think I'd get so emotional just being out here."

❧

When President Roosevelt traveled to Yosemite in 1903, he and John Muir climbed up to Glacier Point one morning and had their portrait taken. The photo is the perfect composite of the two original strains of wilderness appreciation. Roosevelt looks the part of the self-assured outdoorsman: kerchief tied around his neck, jodhpurs and riding boots, a hand on his hip, looking straight into the lens. Muir appears in another world. He stands with his hands behind his back, looking away from the camera, gazing off into the distance.

Combine those two threads into a single twenty-first-century whole and you get National Park Ranger Shelton Johnson. Johnson has been a ranger for all of his adult life. He's been stationed in Yellowstone, in Great Basin National Park and, for nearly twenty years now, Yosemite, where he works as an interpretive guide. Johnson was one of the stars of Ken Burns's PBS documentary about the parks, "America's Best Idea." He stood out with his heartfelt paeans to the wilderness—like when he gets misty-eyed remembering the sound of bison breathing on a winter morning. He has written a novel about the early years of Yosemite, *Gloryland*. He took Oprah Winfrey on her first-ever camping trip. He is one of the park service's most valuable ambassadors.

The day after Chris, Alexia, Michelle, and I wrapped up our trip, I went to see Johnson because I was curious to hear how the wilderness ideal holds up for someone fully aware of the all intellectual critiques. Johnson is a tall, lean, black man, probably in his late forties. When we met at the Native American History Museum in Yosemite Valley, he was wearing the standard ranger uniform—green pants, gray shirt, a wide-brimmed hat—spiked with his own style: a small hoop earring in each ear, a large ring set with amber, a pair of leather bracelets on his left wrist. Johnson is a soft-spoken guy who communicates in a whisper so electrified with enthusiasm that it made the hairs on the back of my neck stand up. When he gets into a reverie, he waves his long, thin fingers in front of him, and it can seem like he's conjuring magic as much as he's describing it.

"Muir said, 'Everything flows,'" Johnson told me as we sat at the base of Yosemite Falls. "When you're in the wilderness, you're in that flow. You wake up when the sun wakes you up, you go to sleep when the stars come out. You can hear the wind blow through the fabric of the tent. You can hear any critter that's out there—whether it's a coyote that's howling away. You can hear that there's not much of a separation between you and it. And the longer you're there, in the wilderness, the separation gets smaller and smaller and smaller. Until pretty soon, there is no separation."

Johnson's Muir-like intensity is so impressive because it's so self-aware. His passion is freighted with history, an acknowledgement of the omissions and the overreaching of the classic image of wilderness. He knows all about the ways in which pre-Columbian societies transformed their environments: "When the first Europeans came here . . . they were actually entering a landscape that had been shaped by people for thousands of years." He recognizes that the idea of wilderness is a uniquely modern construct: "*Wilderness* is a Western term," he told me, "and I have not found in my own studies a corresponding term for *wilderness* in the language of any Indigenous culture anywhere in the world." He's well aware that there is no such thing as a place that's completely pristine: "There are very few landscapes in the world where human beings have never been."

And yet the hope that he places in the power of wilderness remains undimmed. Johnson's appreciation of wilderness is at once postmodern—self-knowing and contradictory—and unabashedly romantic. He has found a way for irony to coexist with righteous sentimentality.

"There are some things that can't be stripped away," he said, as sightseers milled about, some of them eavesdropping on our conversation. "Wilderness is a portal to the earth itself. This is the world, the world as it has been for who knows how many tens of thousands of years. And that's a very powerful transition, to quit thinking of the world as Manhattan, Chicago, Los Angeles. Now we put a line around some area and we call it a 'park.' We call it a 'national forest.' We call it a 'wilderness.' But when you're inside those areas, *that* becomes the world. So now you're on the inside, and the view from the inside of wilderness looking out, wilderness is infinite. Wilderness transcends boundaries. Wilderness only has boundaries when you're on the outside looking in. Wilderness is a place that's bigger when you're inside of it. It's a true wonderland.

"When we're in the wilderness, we're not in Kansas anymore," he said. "I'm not knocking Kansas—I used to live there. But I think I'd rather be in Oz."

Also important for Johnson is the way in which the wilderness continues to serve as an opportunity for self-reinvention. Much of Johnson's career has focused on reviving the history of the California

Buffalo Soldiers, who were among the first federal agents to guard Yosemite National Park. The Buffalo Soldiers are the subject of his novel and the basis of a one-man play he enacts every Sunday evening in the park. It's hard to think of a better example of the transformative power of the wilderness frontier: slaves who became soldiers, tasked with protecting Creation itself.

Johnson's retelling of the Buffalo Soldiers' Yosemite history is his way, he says, of reminding park visitors that "there has always been some kind of diversity here." He told me: "I think it's important to remember that America is basically defined by an immigrant experience, whether that's forced immigration or voluntary immigration. And for much of our experience, it's been an experience of wilderness. We created America out of the wilderness. If we lose wilderness, we'll lose that sense of identity with who we are."

It's an identity, he insists, that anyone can tap into. "If an inner-city kid from Detroit like me can connect to this stuff, anyone can."

What goes down must come up: if you spend three days hiking downriver, and you want to get back to where you left your car, then the laws of physics say you'll have to reclaim the elevation you've lost. Or, as a woman we passed on the trail said, "Once you're in the canyon, you've got no choice but to get back out."

We started early, hoping to beat the heat. But soon we were sweating anyway. The climb was relentless, one steep switchback after another. We trudged heads-down, one foot after the other, up and up, with oxlike patience. The only relief was the fact that our packs were lighter than they had been three days before.

We had covered four miles going a slow and steady pace, Chris hiking in the lead, when we came up yet another rise. I heard him exclaim, "Wow! Now *that's* wilderness." I wondered what we would see. When I got to the lip of the trail and caught the view I had to stifle my surprise. Chris's icon of wilderness was the Hetch Hetchy Reservoir. The wild Tuolumne River that had roared past us for days was now tamed. From above it did look lovely—the water winding around the mountains like a vast cobalt snake. But the river was impounded, and now had all the energy of a bath. "No motorboats, no jet skis," Chris said. "That's wilderness, baby."

In a cemetery in Martinez, California, John Muir rolled in his grave.

We stood there sipping water and taking in the view, and Chris asked about a political campaign he had heard of: "Wasn't there some kind of idea to tear down the dam?"

Here's what I told him:

At the turn of the last century, the fight over damming the Tuolumne River and creating a reservoir in the Hetchy Hetchy Valley was the central environmental struggle of the time. The battle lines were drawn like this: On one side, the California establishment, desperate to find a source of water and electricity for the booming city of San Francisco, then rebuilding from the earthquake of 1906; on the other, John Muir and his newly formed Sierra Club, determined to protect Hetch Hetchy, which they said was every bit as precious as the more famous Yosemite Valley to the south.

The pro-dam forces based their case on utilitarian arguments—a reservoir supplying water and power would serve more people than a wilderness preserve. The reservoir, former San Francisco Mayor James Phelan said, would help "the little children, men and women . . . who swarm the shore of San Francisco Bay" rather than the few people who appreciated "the mere scenic value of the mountains." The highest use of nature was to *use* it, Gifford Pinchot, the founding director of the US Forest Service argued. Pinchot joined the call to dam the Tuolumne, saying that he was all in favor of wilderness preservation "if nothing else were at stake." Of course, something else is *always* at stake. For the defenders of the Hetch Hetchy Valley, that meant the intangible benefits of what they called a "cathedral." Dam opponents tried to push back against the utilitarian arguments, with one Hetch Hetchy defender proclaiming that "dishwashing is not the only use for water, nor lumber for trees, nor pasture for grass."

Since the plan involved damming a river in a national park, the issue had to be decided by the US Congress. After years of political maneuvering and heated campaigning by both sides, Congress in 1913 passed the Raker Act allowing for the construction of a dam within the park's borders. Muir died not long after, supposedly of a broken heart. A century later, the Sierra Club and other environmental organizations continue to call for the dam's decommissioning.

Standing there above the drowned valley, Chris shrugged. To him, the idea of tearing down the dam and restoring Hetch Hetchy Valley was a nonstarter: "Yeah, well, it's a major metropolitan city's water supply we're talking about."

As we started back along the trail, I was troubled. Chris's notion that a city reservoir was some kind of paragon of wilderness seemed to confirm the difficulties caused by what is called the "Shifting Baseline Syndrome." The term—coined by a fisheries scientist charting the ever-shrinking size of wild fish—describes the way in which each generation judges the environment based on the remembered nature of its childhood. The environment you grew up with is what you think of as the "normal" condition of environmental health. But because of the continuing march of human development and the steady degradation of natural systems, that idea of "normal" changes from one generation to the next. What I, as a Gen Xer, remember as pristine forest a baby boomer might see as ruined—not at all like it "once was." In turn, a landscape that I might view as ecologically impoverished would be the norm for a millennial. And onward, until, as a society, we can barely remember what intact wild nature looks like—until what was a horrific desecration for John Muir becomes, for Chris, an ideal of wilderness. When it comes to imagining how wild nature is "supposed" to look, we're running on a hamster wheel of collective amnesia. More evidence, I suppose, of how we all see what we want to see, how each of us apprehends wildness differently.

The lesson hit me again about a mile farther down the trail. We were making our way along a west-facing ridge when we came into a broad stand of aspen. It felt as if we had entered some kind of Arcadia. Thickets of green swaddled the white trunks of the aspen trees—grasses and ferns and sedges in the wetter areas. There were wildflowers everywhere: white yarrow, peach-colored paintbrush, the purple cups of shooting stars. The long, orange, flame-like petals of lilies brushed up against us. Butterflies filled the air—clouds of Boisduval's blue, the black and orange of the Lorquin's admiral, and the western tiger swallowtail's pattern of yellow and black. There was no Romantic grandeur here, just a rich softness that seemed to strike each of us with enchantment.

Out of nowhere Michelle and Alexia became botanical enthusiasts. They broke out their phones and started taking close-up photos of the flowers, giggling as they tried to get the best shot. Chris just stood there, butterflies flitting about his beard, wearing an expression of childlike wonderment. "It's magical, isn't it," Lex said. "It feels like there should be fairies living here."

It *was* magical. A day later, when Shelton Johnson talked of an Oz-like wonderland, this is the scene that flashed through my mind. But then—for me, at least—the spell was broken. I heard the dull roar of a jet tearing through the sky. It was the first engine sound I had heard in days, and with that noise alone I was taken out of Oz and sent back to Kansas. Or, even worse, I was now on my way *flying to* Kansas, stuck in the middle seat, breathing recycled air, munching on peanuts, and watching garbage TV.

No one else seemed to notice. They were all too busy admiring the flowers. But I couldn't get the jet sound out of my head. Even in a wonderland, there was no getting away.

Olympic Wilderness

3

The Forest Primeval

THE RUMBLE OF A JET might be the signature sound of modernity, evidence that we've taken possession of even the sky. The noise is almost inescapable. There have been many times in the backcountry—in the Central Sierra especially, as westbound jets headed for the San Francisco Bay Area begin their decent—when I've had a moment of solitude broken by the dull roar of a jetliner. I'll be at an alpine lakeside, at dusk, remembering that the forest's most common sounds are nothing more than the rustle of wind and the ripple of water. Then I hear the bass thrum of a plane and the moment is over, the wilderness's natural stillness interrupted by an engine.

This wasn't an issue that bothered earlier generations of conservationists. The first commercial jet flight occurred just six years before the Wilderness Act passed in 1964, and the term "jet set" had only recently been coined. Then air travel became commonplace. From 1970 to 1997, the number of jet flights in the United States tripled. Today, if you include commercial jets and small craft, about 8 million planes take off in the United States each year. That's more than 20,000 flights a day.

The clatter of civilization is so commonplace that it has become, literally, background noise. Still, it affects us. The hum of our electronics, the beeping of a freight truck, the never-ending whoosh of cars streaming down the freeway—all of it adds up to a sort of sonic pollution. Health researchers have established that noise pollution disturbs sleep, affects heart functions, reduces productivity, provokes anxiety, and intrudes on cognition. Biologists have shown that humanity's din also messes with the internal harmonies of wild places, the "biophony" as it has been called. Birds can become confused by all of our sounds, making it harder for them to hear their young (though some urban birds have adapted their calls, changing the pitch to compensate for the cacophony of the city). Whales and dolphins are disoriented by the sonar blasts from naval vessels, the drum of freight ships, and the reverberations from underwater oil and gas exploration. Human noises clutter even the soundscape of the deep.

In a little more than a century, we have changed the pitch of the planet. Bernie Kraus, a musician and naturalist who has spent four decades recording the sounds of the biological world, warns that we are acoustically crowding out the sounds of other animals and birds. "A great silence is spreading over the natural world even as the sound of man is becoming deafening," he has said.

All of this racket adds to my feeling of claustrophobia, my concern that the world has shrunk too small. There are, supposedly, just a dozen places left in the continental United States where one can sit for fifteen minutes without hearing a human-created sound. Not one of those places is east of the Mississippi River; you can't find a single such location in all of Europe.

One of the last silent places is the Hoh Rainforest at the center of Olympic National Park, in Washington State. The absence of human sounds there is so profound that, according to an article I had read, the silence feels "like scouring sand." As soon as I heard about such a place, I wanted to go there. I wanted to experience the deep quiet of a place never breached by the noise of chainsaw or bulldozer. So I set out with my lady, Nell, for the farthest corner of the Lower 48 to search out what I imagined would be a kind of sonic oasis.

Getting to such a remote location is no small task. To arrive at the far side of Olympic National Park we had to make a circuitous, five-hour trip from the college town of Bellingham, Washington, through the bustle of islands that make up the Salish Sea—highway to ferry to highway again. Along the way we passed the landmarks that form the working landscape of the Pacific Northwest. Mountain ridges clawed with clear-cuts. Clouds of silver steam rising above the oil refineries in Anacortes, where giant tankers unload crude from Alaska's North Slope. Trucks piled with logs, bound for the pulp mill outside of Port Townsend or the industrial docks at Port Angeles: American trees, destined to be made into furniture in China.

As we went deeper into the timber plantations of spindly, second-growth trees, we began to see signs opposed to a proposed expansion of the park. "No New Wilderness." "No National Park Expansion." "Working Forests = Working Families." A useful reminder, I thought, of how every nature preserve is contentious at some point in its history. Conservation is always a tough call in the moment; only in hindsight does the decision to protect part of the wild world seem obvious.

When we arrived at the Hoh Valley Visitor Center, I asked a park ranger sporting a fantastic walrus mustache about the signs. The center was going to close in five minutes, but he was a generous type, and offered to explain. "That's a big question," he said. "Let's sit down."

We settled onto a sofa in front of a huge window with a view of the surrounding rain forest. "They say the wilderness has no value," the ranger told us. "But look at that parking lot out there. It's full. It's full almost all year round. A lot of people spent a lot of money to come here. And some of that money they spent coming through those logging towns. I think this forest has a lot of value."

I'd agree. Outdoor recreation is a multibillion-dollar business that creates real economic benefits. You could call it the ecosystem service of keeping the tourist dollars flowing. But I didn't travel all that way to explore the financial arguments for wilderness preservation. I was looking for something different in the promised silence. I was going into the deep forest to explore tougher questions. Does a place still have value even if it is of no obvious use to humans? What's a wild place worth for everything that doesn't walk on two legs?

A rainforest is the least welcoming wilderness. We humans seem predisposed to appreciate open spaces: meadows, prairie, the kind of parklike woodland in which the trees are spaced widely. This is probably an evolutionary tic. We prefer those places that remind us of the African savannah, a setting that is not just the birthplace of consciousness but also has real, physical advantages. On an open plain you can see the predators coming and you can spot where the prey is going. The oak-studded grasslands of California are a good example, as are the rolling hills of Britain's Lake District, so beloved by the Romantics. Even the desert—the fearsome wilderness of the Bible—boasts an open vista.

The forest is altogether different. It's dark, close, and shadowed. There is a chance that some terrifying thing may be lurking just beyond sight. The word *savage*—with all of its negative connotations of unchecked wildness—comes from the French word *sauvage*, which in turn is derived from the Latin *silva*: "forest." Whether this fear of the forest is learned or innate I'll leave to others to debate. All I know is that it runs deep. Once, in the course of doing some reporting on the tropical rainforests of Indonesia, I had an environmentalist ask me not to use the word *jungle*. When people hear that word, I was told, they think of a scary place.

The closest you can come to a jungle in the United States outside of Alaska or Hawaii is the temperate rainforest on the windward side of the Olympic Mountains. In an average year, 142 inches of rain fall on the Hoh River Valley—nearly twelve feet of precipitation. The nearby community of Forks is so relentlessly overcast that it was the setting for the *Twilight* series, and by now the scrappy log-ging town has made a cottage industry of hosting "vampire tours."

Nell and I hadn't hiked more than a few miles from the trail-head when the forest gloom started to affect us. "This is the most wilderness-y wilderness you've ever taken me to," she said as we hiked upriver, the slate-gray waters of the Hoh rumbling along-side of us. You should know that this is a woman who has kayaked through rainstorms, spent icy nights in the desert, and climbed up and over mountains too numerous to name. "It's the forest," she said. "It's spooky here. There's an eeriness."

I knew what she meant. It felt like we had entered the forest primeval, a forest so ancient that it seemed more geologic than bio-logic. The bottomlands along the Hoh Valley are a mix of hemlock and spruce sprinkled with red cedar and bigleaf maple. In the fertile flats along the river the trees are enormous, many of them well over 100 feet tall. From the high canopy, life cascades downward in lay-ers; a drop of rain can smack into as many as twenty different levels of leaf, needle, and lichen before hitting the forest floor. There's a musty, mildewed smell in the air, the scent of millennia of accumu-lated leaf mold.

All of the dozens of kinds of mosses deepen the mood. The moss is everywhere, hanging in great, gray sheets from tree limbs, shroud-ing the downed logs in green. Moss on rock, moss on wood, tumor-ous, creeping. The long fingers of moss shift with the lightest breeze, multiplying the shadows and casting them deeper.

The hemlock trees are witchy as well. The trees bend at their tips, and the droop of their branches, cloaked in moss and liverworts, make them seem as if they are in the midst of decay. The Hoh is a world of rot: the downed logs seeping into the soil, the suffocat-ing moss, each living thing teetering on the edge between life and death. It seemed the kind of place where filmmaker Tim Burton would like to have a spooky little cottage back in the woods.

In that alien landscape my first task was to pay attention, to bend all of my energies to noticing the more-than-human world. Hiking the Grand Canyon of the Tuolumne with Chris and Alexia, I was looking for the original anthropocentric values of wildness—why it's good for us. Here in the rainforest, I was looking for the *biocentric* values of wildness—why it's necessary for all the other life forms we share the planet with. To make that leap to considering the interests of nonhuman life, before anything else I had to take on the eye and the ear of the naturalist. That is, I needed the patience to pay careful attention.

Encouraging a close attention to wild nature has been one of the main currents of American environmentalism. In the late nineteenth century, one of the few conservationists who could compete with John Muir in public stature was a best-selling author named John Burroughs. In one of his classic essays, "The Art of Seeing Things," Burroughs wrote, "I take pleasure in noting the minute things about me. I am interested even in the ways of the wild bees, and in all the little dramas and tragedies that occur in field and wood." A century later, Annie Dillard won the Pulitzer Prize for a similar meditation on the wonders of "minute things." In Dillard's *Pilgrim at Tinker Creek*, the excitement with wild nature doesn't happen in some majestic setting like Yosemite, but rather in the trashed-out wood-lots of Virginia, prompted by acts as simple as stalking muskrats. The wonders of creation are all around, Dillard insisted. "It's all a matter of keeping my eyes open," she wrote. "Nature is like one of those line drawings that are puzzles for children: Can you find hidden in the leaves a duck, a house, a boy, a bucket, a zebra, and a boot?"

Once you begin to pay close attention to life's details, new layers appear. The landscape becomes readable. Walking along the banks of the Hoh, I was beginning to see the patterns of the place. I noticed that the alders were all clustered beside the river-bank. I could see the more widely spaced spruce groves revealing where the elk graze. I nibbled on the licorice fern growing off the trunks of the big leaf maples, just as the Hoh Indians once did. Yet I knew I was still just touching the surface. It takes practice to have the naturalist's eye. One must know how to see, but also what to look for.

Yet you can go too far with this. Different kinds of knowledge shape different types of understanding, and an obsession with nam-ing can end up narrowing perspective. The novice becomes the expert, who then becomes the specialist who puts on blinders in order to focus better. "To see the scarlet oak, the scarlet oak must, in a sense, be in your eye when you go forth," Thoreau wrote. "We cannot see anything until we are possessed of the idea of it, and then we can hardly see anything else."

So it's best to keep some of the beginner's mind. God forbid that I should ever know the names of all the plants and flowers in any forest. Sometimes I like not knowing exactly what I'm seeing: as long as you have to keep guessing, you can stay wonderstruck. As Nell and I continued upriver it was good fun to debate which ferns were the lacy lady ferns and which were the lacy maidenhair ferns.

We cannot see anything until we are possessed of the idea of it. That sentence alone confirms the claim that wilderness is a human construct. Was the forest really eerie, or had we just learned to see it that way? Yes—both. There's no question that I was seeing the forest through the veil of my own desires and fears. And there was no question that the wildness was implacably real.

With the naturalist's eye I noticed the primal struggle of survival all around me. In that ancient forest the elegant economy of death was apparent. The trees, I could see, were often arranged in straight lines. Odd as it may sound, the orderliness was evidence of the brutality of competition and the way that death begets life. The sopping-wet forest floor is too dank for a tree to take root; the seeds just rot. Usually the only spot where a tree can sprout is on the top of fallen trees, which are called "nursery logs." In the trees' adolescent stage of life, the arrangement is obvious: a half dozen saplings growing along the centerline of a downed tree. Then some twenty or thirty years pass, the nursery log fades into soil, and all that's left are parallel lines of spruce organized in a perfect arcade, the trees birthed and fed from the corpses of their grandparents.

The Hoh's inhospitableness to even its largest inhabitants seemed an exact illustration of Charles Darwin's claim that life is constantly straining against other life in a "struggle for existence" in which "a grain in the balance will determine which individual will live and which shall die." The author of *The Origin of the Species* is key here. Because when we understand Darwin's insights about the nature of life, we gain an appreciation for the intrinsic value of wildness— wildness for the sake of wildness.

If you accept that evolution is life—that life's wondrous diversity is formed out of the processes of natural selection—then suddenly Thoreau's koan that "in wildness is the preservation of the world" becomes more understandable. The line is literally, physically true. Wildness preserves evolution. And evolution, the daily audacity of creation unfolding, keeps open the possibility of new forms of life.

When Darwin is included in the story of wilderness, it puts to rest the notion that preserving wild places is somehow a nostalgic endeavor. Just the opposite. Protecting some parts of the world's wildness is about protecting the future. Or, more importantly, the *futures*—the very possibility of unimagined possibilities.

Darwin's insights about life expand our imaginations, and that imaginative expansion can also lead to a moral expansion: the thought that maybe, just maybe, the rest of life has the same rights as we do. The ferns, the trees, the vagrant shrew that lives in the Hoh rainforest understory—each of them possesses the same inalienable rights as you and I. They also deserve the rights to life, liberty, and the pursuit of happiness. Or, in the case of the Pacific fisher (an endangered, weasel-like critter in the Hoh) the right to pursue and eat those shrews.

Those are radical ideas, I know. It can be hard to imagine that a dumb shrub should have the same rights as a being that can construct abstract ideas like "wilderness." But I wasn't the first person to have such crazy ideas there on the banks of the Hoh River.

They called him "Wild Bill." It's an unlikely handle for a US Supreme Court Justice, but William O. Douglas earned it fairly. The nickname partly had to do with his personal indiscretions. Douglas was a notorious womanizer who went through wives the way most justices go through law clerks. Douglas kept getting older and older, but his wives (four by the end of his life) stayed the same age; that is, they were most often in their twenties when he married them.

The nickname also alluded to the fact that Justice Douglas was an avid outdoorsman who affected the rugged mannerisms of a Westerner. When the Supreme Court was not in session, Douglas was usually out hiking and fishing in the country's wildest areas, East and West. Over the course of several of his thirty books, the justice shared his passion for wilderness with his fellow citizens. "Mountains can transform men," he wrote in a uniquely American style that combined Jeffersonian deism with an alpine fetish. "When man ventures into the wilderness, climbs the ridges, and sleeps in the forest, he comes in close communion with his Creator."

Douglas found this wilderness faith early. He grew up in Yakima, Washington, in the dry foothills of the eastern Cascades. As a child he suffered from infantile paralysis, which, his mother feared, would leave him crippled. The boy wouldn't have it. Driven by a fierce will, the young Douglas set off for the nearby mountains whenever he could. He built up his body—and soon his spirit and intellect—by exploring the wild. In the wilderness, he later wrote, a person "may find their own relationship to the universe in the song of the willow thrush at dusk."

Douglas was a hardcore angler, and one of his favorite places to fish was in the salmon and steelhead runs of the Hoh River. Going deep into the Olympic Mountains, Douglas wrote, "the roar of civilization is left behind," and one can find "a sanctuary where voices above a whisper seem almost sacrilegious." In the rainforest valleys of the Hoh and the Quinault, the justice discovered "an important lesson in ecology.... No species should ever be eliminated, for man in his wisdom does not yet know the full wonders and details of the cosmic scheme."

Douglas brought this backwoods ethic to the Supreme Court, where, in addition to being a stalwart defender of civil liberties, he was the most energetic tribune of the natural world in the court's history. You can think of Douglas as, say, the Thurgood Marshall or Louis Brandeis of ecological liberties.

Justice Douglas's most important articulation of the rights of nature came in a dissent he wrote in the case of *Sierra Club v. Morton*. In the early seventies, the Walt Disney Corporation was making plans to develop a ski resort in a place called "Mineral King" adjacent to Sequoia National Park. The Sierra Club opposed the resort, and filed a lawsuit. In 1972, the case made its way to the Supreme Court, where the dispute rested on the question of whether the environmental group had the legal standing to sue on behalf of an ecosystem.

In the end, the case was something of a wash. A majority ruled that the Sierra Club did not have standing to bring a lawsuit; Disney ended up dropping its development plans; and the area was eventually absorbed into the national park. (The case ended in an unusual four-to-three decision, as two justices recused themselves and each of the three dissenting justices wrote their own individual opinions.) But the case had a lasting influence because of Justice Douglas's bold declaration that, regardless of whether or not the Sierra Club had legal standing, the forest itself should enjoy the protection of the law. The case, Douglas wrote, should more properly be called *Mineral King v. Morton*, because it was the land that was facing injury. If a corporation—"a creature of ecclesiastical law," in Douglas' words— had rights, then why not a river? The justice went on to write: "The river, for example, is the living symbol of all the life it sustains or nourishes—fish, aquatic insects, water ouzels, otter, fisher, deer, elk, bear, and all other animals, including man, who are dependent on it or who enjoy it for its sight, its sound, or its life. The river as plaintiff speaks for the ecological unit of life that is part of it. . . . The voice of the inanimate object, therefore, should not be stilled."

Douglas's opinion was a minority of one; no other justice signed his dissent. So it has often been. A majority of people have always found it difficult to imagine that the rest of the planet should have the same rights as people. "Each time there is a movement to confer rights onto some new entity the proposal is bound to sound odd or frightening or laughable," Christopher Stone, a young legal scholar, wrote in a law review article titled "Should Trees Have Standing?," which Justice Douglas cited in his ruling. In his essay, Stone acknowledged the historical, philosophical, and emotional difficulties that any expansion of rights entails: "Until the rightless thing receives its rights, we cannot see it as anything but a *thing* for the use of 'us.' . . . Such is the way the slaveholding South looked upon African-Americans."

Although the rights-of-nature idea might have seemed "laughable" to some people, the arguments of Douglas and Stone fit perfectly with the counterculture of the time, an era in which an emerging ecological consciousness was dramatically reshaping American law. The early 1970s saw congressional passage of the Clean Water Act, the Clean Air Act, the National Environmental Protection Act, the Marine Mammal Protection Act, and the Endangered Species Act. Of these, the Endangered Species Act, passed in 1973, went the furthest in articulating the rights of other living beings. For starters, the law covers not just mammals, but also fish, amphibians, insects, and plants. Under the definition of the law, harming a species means not just killing it, but also significantly disrupting the habitat it relies on, a provision that comes close to acknowledging the rights of whole ecosystems.

The Endangered Species Act and the Wilderness Act were the highest legal expression of an environmental ethic that had been forming for some time. In the first half of twentieth century, conservationists were starting to articulate a coherent philosophy of humans' moral responsibilities to the rest of creation. Few voices were as influential in articulating that idea as Aldo Leopold.

In the pantheon of environmental thinkers, after Muir comes Leopold. Aldo Leopold was born in 1887 to a well-to-do family in the farming community of Burlington, Iowa. He grew up hunting and bird-watching on the banks of the Mississippi River, and later he took his enthusiasm for the outdoors to the then-new Yale School of Forestry. At Yale, Leopold internalized the utilitarian view of natural resources taught by Gifford Pinchot. Soon he joined the US Forest Service. In his early writings, Leopold, like conservationists before him, emphasized the wild's benefits to humans. "I am glad I shall never be young without wild country to be young in," he wrote. Working within the Forest Service bureaucracy, Leopold fought for the creation of the first designated wilderness areas—"primitive areas," the Forest Service called them then—forty years before the passage of the Wilderness Act. By *wilderness* Leopold meant "a continuous stretch of country preserved in its natural state, open to lawful hunting and fishing, big enough to absorb a two weeks' pack trip, and kept devoid of roads, artificial trails, cottages, or other works of man."

Over the years, as he spent more time in wild nature, Leopold's thinking deepened. The forester became a philosopher, one who believed that humans had ethical responsibilities to nonhuman life. Leopold's career coincided with the emergence of ecology as a scientific discipline, and perhaps his signature accomplishment was the way in which he welded morality to science, in the process formulating a new way of thinking about humanity's relationship to wild nature. "That land is a community is the basic concept of ecology," he wrote in the foreword to his seminal book, *A Sand County Almanac*. "But that land is to be loved and respected is an extension of ethics."

Leopold called this, simply, "the land ethic." There is, he said, a "community of life" that includes "soils, waters, plants, and animals, or collectively: the land." This "land-community" has, at the very least, a "right to continued existence." When we recognize this right, we can make clear judgments about whether our actions toward nature are right or wrong. Leopold wrote: "A thing is right when it tends to preserve the integrity, stability, and beauty of the biotic community. It is wrong when it tends otherwise."

The land ethic marked a huge departure from how humans previously thought about morality. More to the point, it represented an *expansion*. As Leopold noted, ethics originally involved relations between individuals—the Judeo-Christian Ten Commandments being a classic example. Then ethics expanded to include the relationships between individuals and society—the questions of politics that preoccupied Socrates and Plato. It is now time, Leopold argued, to "enlarge the boundaries of community" to include the nonhuman.

Or, as I have always thought of it: There are two main ethical questions that humans of compassion must grapple with. The first—the one that has consumed moral philosophers for most of history—is how to treat other people. What are the requirements of justice and equality and liberty? The second—the one that Leopold articulated so clearly—is how to treat *other species*. What are the requirements of *ecological* justice and fairness? Whether we are able to show some restraint in our actions to safeguard the rights of the millions of other species on the planet is also a test of our ethical mettle.

Leopold didn't live to experience his influence. His small masterpiece, *A Sand County Almanac*, was published posthumously, and at first was misunderstood as merely a collection of quaint nature essays. It sold only a few thousand copies. Not until the 1960s did Leopold's environmental ethic resonate broadly. Wallace Stegner called *A Sand County Almanac* "one of the most prophetic books, the utterance of an American Isaiah."

Leopold probably wouldn't have been surprised by the delay. He recognized that ethics are "a product of social evolution," and that they take time to develop. And, in fact, his ideas have continued to evolve since his passing. In 1975, philosopher Peter Singer published *Animal Liberation*, which decried "the tyranny of human over nonhuman animals" and laid out a careful argument for individual animal rights. Around the same time, a Norwegian mountaineer, Arne Naess, coined the term *deep ecology*: a philosophy of ecological egalitarianism "wherein there are no sharp breaks between self and the other." In recent years, some environmentalists have pushed for legally binding "rights of nature" laws that recognize the intrinsic rights of whole ecosystems.

These ideas exist along a spectrum. While Singer makes a claim for the intrinsic rights of individual animals, the nascent rights-of-nature movement permits the killing of animals for food and clothing, as long as the integrity of biotic communities is maintained. But all owe a debt to Leopold's legacy. "In short," he wrote, "a land ethic changes the role of *Homo sapiens* from conqueror of the land-community to plain member and citizen of it."

We humans are not the rulers of Earth. We are, simply, neighbors with the rest of life. More accurately, we're roommates sharing the same home—a one-room house floating in the middle of space.

Sitting on the banks of the Hoh River, reading my Douglas and rereading my Leopold, it seemed to me that the wild is so important because it's leveling. It makes us see that we don't deserve to be as high and mighty as we often feel. Yes, sure, we have the power to assert our will whenever and wherever we like. As a uniquely powerful species, we can do as we please. But there's no morality in that, no charity and no honor. To act however you like without regard for others—that's just the behavior of the tyrant or the sociopath.

In one of his essays, Leopold urged people to think like a mountain. I love that—*thinking like a mountain*. But I'll admit that it can feel impossible to take on a geologic view of things, to widen my empathy that far in space and time. The only way to come even close to thinking like a mountain is to be sitting on or next to one. As Leopold put it, the wilderness "builds receptivity" to the idea that we owe moral obligations to other beings. It's the kind of radical idea that you have to see in order to believe.

❦

Nell and I had covered a decent bit of ground the first day in the forest: eleven easy miles along the river's edge, which we took at a gentle pace. Now, to get up onto the shoulders of Mount Olympus, we would have to work. As we climbed out of the rainforest and into the beginnings of the montane ecosystem, the landscape changed. The trees spread apart and great pastures of sword fern carpeted the halls of spruce and fir. As we hiked, our footfalls echoed through the perfect symmetry of the groves.

Flowers appeared. There had been few flowers in the depths of the rainforest, but now, as we climbed into the air and the light, color arrived on the scene: bright pink fireweed in the rocky areas, the big white umbrella of cow parsnip and billows of leatherleaf saxifrage in the cuts. It was as if altitude had channeled time, and we had entered a miniature version of the Cretaceous explosion—that moment some 130 million years ago when, without warning, flowers burst into being.

At one point the trail went along a steep ledge where the mountainside dropped off sharply to the right. Suddenly we were hiking among the treetops. The massive firs were rooted somewhere down below—even craning over the ledge we couldn't see their base—and then rose up a hundred feet and more until the tree crowns met us at eye level. And there we discovered that the arboreal heights were full of life. Vast clouds of tiny white moths fluttered among the needles and cones, tens of thousands of moths, probably more.

Later, checking my field guide, I would learn that the creatures are called "pine whites." They lay their eggs at the highest reaches of conifer trees and spend most of their lives there. In the moment, though, unequipped with any scientific knowledge, I found the sight spellbinding. All of the air was aflutter, the whole scene was blinking, and my eyes couldn't keep up. It felt like the day had been put under a giant strobe light. The swirl of moths was so thick and so fast that the image of the world quavered, as if nature's signal were coming through too strongly and had overloaded the picture.

We stood and looked for a good couple of minutes—just looked, with unchecked wonder. Then it hit me: these small white moths had been here all along, even when we were unaware of them. Most of the time, their tiny niche in the world goes unnoticed by humans. They can spend their whole lives beyond our gaze: hatching, breeding, eating, dying far above our heads. I thought of a line from Thoreau: "What we call wildness is a civilization other than our own." I thought of Edward Abbey, his observation that the wilderness is "a realm beyond the human." And I remembered a moment from the Yosemite trip: on the last day we had been pumping water from a small stream when Chris spotted a trout in tiny pool, and was shocked by its mere being. "What do you *do?*" he wondered with eyes wide open. Thus does the wild stun us with the reminder that most of existence occurs without anyone paying attention.

"Most members of the land community have no economic value," Leopold wrote. "Wildflowers and songbirds are examples." As are pine whites. We can live our lives quite comfortably without them—without even knowing they exist. They could disappear from the face of Earth and their demise would likely go unheralded.

But a species' instrumental value to humans shouldn't matter. Those pine whites have an *intrinsic* value. They have a worth in and of themselves, even if it appears that they're good for nothing. For a being to live independent of humans is, in some ways, the essence of wildness. The pine white lays its eggs in the tops of fir trees. It flits about among the needles and branches. We do not care. We do not even notice. And so the moth demonstrates its self-will. Unlike the hog or the broccoli or the hybrid tea rose, the pine white isn't embroiled in any dance of domestication with us. The moth exists without our permission.

Take a minute to consider the interests of a moth, and it becomes evident that *every* landscape is a working landscape. It's just that the landscape may not be working for us. It's working for itself—for the continuation of evolution. No matter how small a being's ecological niche, that being has a right to continue living, if only to serve the greater good of evolution unfolding.

This doesn't mean that every species has a right to exist forever. Evolution's a bitch, and extinction is the unavoidable fate for most everything on the planet, including us. But each species possesses at least the right not to be snuffed out by humans.

Unfortunately, on that score we're not doing very well. Biologists warn us that we are in the midst of the planet's sixth mass extinction. On five other occasions in the history of life on Earth, some dramatic event—a sudden spike in CO_2 levels, an asteroid hitting the planet—has caused a massive die-off of species. Today a similar die-off is occurring. Scientists estimate that a quarter of all mammals, a fifth of all reptiles, and a sixth of all birds are on their way to oblivion, percentages that are much bigger than the historical average. Amphibians are going extinct at a rate 45,000 times higher than the historical rate.

We humans are the primary cause of the disappearances. Our ever-spreading development chews up wildlife habitat, while the pollution from our factories and cars heats up the planet and causes further dislocations. It's not just the scale of the changes we are making on Earth's systems, but also the speed at which we are doing so, a phenomenon that has been called "The Great Acceleration." Our technologies outpace species' abilities to respond and react. According to journalist Elizabeth Kolbert, author of *The Sixth Mass Extinction*, "now we're the asteroid."

If any stratigraphers are examining the fossil record eons from now, the abrupt disappearance of biodiversity will be among the clearest signs of the Anthropocene. The impoverished fossil record will show that we humans had become the greatest evolutionary force on the planet. We're sort of like a one-trick Shiva, practiced only at destruction.

To destroy a species forever is a crime; to do so thoughtlessly makes it something closer to a sin. Of course, it's impossible to mourn what you do not know. The wilderness is important, then, because it reminds us of the all the life we normally never see. Pine whites, for example: nothing more than tiny, little moths that can enlarge the scope of our sympathies.

That evening Nell and I made it up to the shoulders of Mount Olympus. The base camp for the climbers preparing to bag the peak is called "Glacier Meadows," and I had imagined an alpine idyll, a broad sweep of grasses capped with a postcard view of the summit. To my disappointment, it was nothing like that. The campsite was set in a thick wood of alpine firs, smaller and tougher than their downslope cousins, and there was no view at all, just the trees packed close and tight, relieved only by a narrow streambed to the west. The space felt cramped and claustrophobic.

As night closed in, though, I found what I had come looking for—a deep silence, as deep as a dark well bottom. But the silence also came with a feeling of disquiet. The place felt spiked with an unsettling loneliness.

In September 1937, President Franklin Delano Roosevelt visited the Olympic Peninsula and in a speech there called on Congress to create a national park that would serve as "a great pleasure ground" for Americans. *A great pleasure ground*—that pretty well sums up the view of nature that has guided the park ideal since Yosemite Valley was protected more than 150 years ago. Preserves of wild nature are supposed to give us joy and delight. But that's only part of the story. The wilderness at its wildest isn't always about fun. Much of the virtue of the true wild comes from experiencing fear. Any foray into the wild should have an element of danger and risk. If you're not feeling just a little bit afraid when you enter the wilderness, then you're doing it wrong or not paying close enough attention.

Now, in the fireless night, Nell admitted that she was scared—and in fact had been for much of the last two days. "There are places where even the native peoples wouldn't go, and for some reason we rush to go there," she said as we sat in the dark, sipping tea. Mostly, she was afraid of wild animals—the bears that, though we had not seen them, we knew were roaming the forest. "I'm afraid we'll see a bear, and I won't know what to do, and that I'll get mauled. That I'll get a claw in the kidney."

Nell had good reason to be worried. There aren't any grizzlies in the Hoh, but there are black bears, and even the smaller species of *ursus* isn't an animal to take lightly. A claw in the kidney would, indeed, mess you up bad. Which is why, as soon as you enter a wilderness that includes large predators, the mood changes. If there is any one thing that distinguishes the intermittent wildness of the nearby nature with the immersive wildness of the remote wilderness, it's the fear sparked by the presence of large predators. Or, if not fear, then at least a heightened sense of alertness. I've done a lot of solo backpacking, and whenever I'm in cougar territory, especially, I carry myself warily. My hearing becomes sharper and my sight becomes more attuned. My skin feels more sensitive. It's as if a sixth sense kicks in: the instinct of being on guard. I know it's a cliché, but in the wild you really do feel more in touch and alive—if only because that's the key to *staying* alive.

As I've said, the pastoral landscape of the garden or the parkland has a lot to offer us: the play of the seasons, the surprises of phenology as we track the comings and goings of birds and flowers and in doing so hitch ourselves closer to the current of life. The charms of the woodlot, however, are usually missing the element of violence (at least, violence on a scale humans can notice and appreciate). Our state parks, regional preserves, and the front country of the national parks can often feel like a diorama: nature stilled and presented within a frame. There's birdsong and plant life, but nothing menacing. The daily cadence of death—the churn of fang and claw—is missing. As soon as bears or mountain lions or wolves enter the picture, you're no longer merely admiring the scenery. Suddenly you're in the middle of the action. As poet Gary Snyder says, the predator's presence "is ecology on the level where it counts."

Unlike the alertness we feel in the cities—where we're mostly on guard for a wayward car or an out-of-control person—the fear we experience in the wild is a productive fear. Large predators like bears and mountain lions put us in our place: a spot just a notch below the apex of the food web. This is especially important in the epoch of the Anthropocene, as some people blithely assure us that, because we have an opposable thumb and a nice bit of gray matter, we're entitled to put our mark on every landscape. Being among big carnivores is a reminder that—even though humans are what biologists would call a "generalist species"—we, too, have our niche, and it's not always right at the top.

Nell and I didn't see any bears, but earlier that evening we did happen on a pair of mountain goats. A mama and an adorable kid, both with incredibly pure white coats. They seemed harmless, and went along munching the grasses as we looked for a tent site. If anything, they were too habituated to humans. I knew, though, that the thick black horns on the mother's head could be dangerous.

Just a few years earlier, one of those mountain goats had killed a man from Port Townsend. He was hiking on Hurricane Ridge, a popular trail at the north end of the park, when a ram charged him. The goat's sharp horn took him straight in the thigh and sliced into his femoral artery. The hiker's companions tried to scare the beast away. They yelled and threw rocks and used a picnic blanket as a whip. But the goat wouldn't give ground. The animal stood over the man as he bled out on the side of the path.

It's worth noting that the mountain goats in the Olympic Mountains aren't native to the area. They were introduced in the 1920s to serve as hunting game. Depending upon whatever baseline you have in mind, they might not be considered "natural" to the area, though they are most certainly feral—evidence that a place doesn't need to be pristine in order to be wild.

The next morning we hiked up to the ridges above Blue Glacier and stayed there awhile admiring the soft-serve-like mound of snow atop Mount Olympus. We're no climbers, however, and had no intention of reaching the summit. So we headed back the way we had come, thousands of feet down, past the picture-perfect scene at Elk Lake, past a staircase of waterfalls, and then once again into the rainforest with its mosses and its musty smell.

Between the climb up to the glacier and the trek back down into the valley, it ended up being a long day. By the time we got to the camp marked as "Olympus" on the map and found a good spot on a sandbar alongside the river, we were done for. I made a quick dinner and Nell crashed out, leaving me with the ramble of my thoughts.

Bears. Butterflies. The challenge of holding onto a biocentric worldview. I kept thinking about the far side of the river—the groves where human consciousness never goes. Does a storm of pine whites still make the treetops quiver even if there's no one around to see it?

Thinking like a mountain, Leopold said. At best, such an ideal leads to a generosity of spirit. At the very least, it can prompt a recognition of our shared interests with the wild. When we see that our fate is intertwined with that of other species, our concern for nature goes from mere *noblesse oblige* to true ecological solidarity. Human self-interest isn't incompatible with biocentrism; it just proves Darwin's point that everything strives for its own survival. When we enter into ecological solidarity, we humans are just being good animals, protecting our own interest in order to thrive.

I suppose one can learn that lesson from a book. The best instructor, though, is the wild itself. "A species whose technological cleverness has made it the schoolyard bully desperately needs the ethical discipline that wilderness provides," Roderick Frazier Nash has written. "Wilderness is the best place both to learn and to express ecological limitation."

In the gloaming I went back into the woods to hang our food bag out of the reach of bears. As I was returning to our tent site on the riverbank I heard the strangest sound.

It was coming from upriver, from the end of the valley, where a fat, orange moon was rising above the eastern peaks, now swaddled in wisps of cloud. A low hum, almost like a jet engine, but then above that a slapping bass sound, like a fan whapping the air. Whatever it was, it was low to the ground. I stood still to listen. It came closer and the vibration got louder. Then it became clear—an engine and helicopter props. And then it was upon me—a twin-propeller Chinook helicopter, flying just a couple of hundred feet above the river, the huge black shape lit with a few red and blue lights on its body, the engine noise ripping apart the ripple of the river.

I stood on the sandbar alone and waited for the river murmur to again become the loudest sound around. But the scene wasn't the same. For a few days I had been able to imagine that I inhabited the wild animals' world. As the helicopter racket passed down the valley, it became clear what a fantasy that had been: in fact, the animals live in ours.

Arctic National Wildlife Refuge

4

Fall of the Wild?

FINALLY, I DISCOVERED THE UNINTERRUPTED WILD I had been searching for—though I had to travel to one of the farthest ends of Earth to find it.

For five days we had paddled the Aichilik River through the Arctic National Wildlife Refuge, the fabled wilderness on the north slope of Alaska's Brooks Range. Now, we were at the shores of the Arctic Ocean. After a few of us took a plunge into the ice-filled waters (having come all this way, why *not?*), the group settled in for the everlasting evening. Most everyone had gone to sleep, but I decided to stay up, determined to catch every last minute at the top of the globe.

The stillness was flawless. There was no wind—nor leaf or grass for wind to stir, anyway—and the water was perfectly flat, unblemished, like a plate of brass. As the sun made a lazy arc through the northern sky the temperature dropped into the forties, sending a fog off the ocean. The light turned thick as honey; sea and sky fused into a single field of orange. An immense silence descended. Now and again I could hear the muffled boom of an ice sheet collapsing off in the distance, the wing beats of eider flocks zipping among the ice floes.

I would call the scene all pristine—but I know better. *Ethereal* was more like it. Otherwordly. Here remained a place where one could escape the sights and sounds of civilization. And here, too, was a place large enough and remote enough for evolution to rule. The Arctic, at least, remained the wild of the imagination.

And yet. The things we had experienced during our trip down the Aichilik had only deepened my doubts about the future of wildness. The tundra was untamed, but it was not untouched. Even in such a remote place, civilization's thumb pressed firmly on the scale. Worrisome new questions had arisen in my mind during the course of our group's float down the river. In this Human Age—with the force and speed of our technologies causing unprecedented ecosystem damages—was it still worthwhile to try and keep some places wild? Especially if such a hands-off approach might doom plants and animals to extinction? Don't we have a responsibility to make an effort to repair the damage we are causing, though that would likely mean trammeling the wilderness?

The thought was disturbing: in a time of environmental hardship, maybe the wilderness has become a luxury. If we want to help nonhuman nature survive the global fever we've created, we might have no choice but to bring wild landscapes further under human control. To "save nature," will we have to sacrifice wilderness?

We had started from Fairbanks a week earlier. The expedition had been put together by Dan Ritzman, the head of the Sierra Club's Alaska program. For nineteen years Ritzman has spent a portion of each summer guiding rafting trips down the rivers of the North Slope. He has taken celebrities and US senators and other civic leaders out onto the tundra as part of the larger campaign to prevent oil drilling in the wildlife refuge. Ritzman's love of the vast, open spaces of the Arctic runs deep ("As a Sierra Club staffer, I probably shouldn't say this," he joked at one point, "but I don't really like trees"), and he has a zeal to share the place with others. Our trip down the Aichilik River would be the latest installment of that public education tactic. Ritzman had recruited an impressive crew including some of the Sierra Club leadership, an artist, an activist, and a photographer, all of whom, it was hoped, would go home and spread the word about the wonders of the Far North.

Probably the biggest "name" on the trip was Paul D. Miller, aka DJ Spooky. Spooky came up in the Manhattan club scene of the late nineties, a trip-hop turntablist who was pals with the hit-maker Moby. He then branched out into more avant-garde concerto composing and multimedia installations. He had performed at the Metropolitan Museum of Art in New York, had been included in the Venice Biennial and the Whitney Biennial. Miller's latest passion is global warming's impact on polar regions (he had been touring a performance that mixed a string quartet and film footage of a melting Antarctica), and Ritzman figured the DJ would be an influential ambassador for the cause of Arctic protection. Miller is adventurous in his own way—he had spent time at a research station in Antarctica and had been tapped as a National Geographic Society "emerging explorer"—yet he seemed a bit underprepared for the adventure. In his Sta-Prest khakis, tight denim jacket, and white newsboy cap turned at a rakish angle (an outfit he would maintain the whole time on the tundra), he looked like he was going out for a night at the club, not a week in the wilderness.

Ritzman had also recruited Rue Mapp, a social entrepreneur from Oakland, California, who had started an organization with the mission-explicit name Outdoor Afro. Mapp had distinguished herself as an up-and-comer in the environmental movement—she had been invited to the White House, that sort of stuff—and it was clear to me that the Sierra Club was grooming her for bigger things. For her part, Rue was eager to explore new links between the environmental and racial justice movements. "I think it's interesting that the Civil Rights Act and the Wilderness Act were both passed in the same year," she had said during a welcome dinner at our hotel in

Fairbanks. "A big opportunity was missed then to connect the two. But we have this chance now to articulate how access to wild places is also a civil rights issue. When you're in the wilderness, it just opens up your sense of possibilities."

The other VIP was Michael Brune, the Sierra Club's relatively new executive director. Brune had made a name for himself fighting coal companies and getting hauled out of international environmental summits for staging protests. But he didn't have much experience in traditional conservation issues, and the trip had been organized, in part, to give the new boss a dose of that old-time wilderness religion.

I should say now, by way of disclosure, that I've known Brune for years and consider him a friend. That's how I had sneaked onto the trip. Brune told me about it at a Christmas party and we hatched a plan to go together. Brune (which is how everyone I know refers to him) is a tall man, probably six-foot-four, with that rare gift of projecting his presence when he enters a room. Yet he's careful about not taking up too much space. He listens more than he talks, and when he does it's usually with a wry humor honed to disarm. Though he has lived in California for years, he still has the bluntness and relish for battle of someone born and raised in New Jersey. One of the best parts of his job, he told me during one of our strolls on the tundra, is the chance to be "politely rude" toward those in power—that is, going to meetings on Capitol Hill or in the West Wing and being a charming pain in the ass in defense of the environment.

Also along on the trip was Shirley Weese Young, a Sierra Club board member from Chicago who had been invited along because, as she explained, the other board members knew she could hack it. Now in her sixties, as a younger woman Shirley had sailed across the North Atlantic as well as through all of the Great Lakes, and there was a certain daring behind her sparkling eyes. A young photographer-videographer from Boulder, Micah Baird, would accompany us to capture the experience. A river guide—a part-time biologist named Peter Elstner who spends the winters piloting drones around the Arctic to collect wildlife data—was already camped out in the foothills of the Brooks Range, waiting for us.

It was only a couple of days after the summer solstice, and as we drove from our hotel to the Fairbanks airport for a 7:00 a.m. takeoff, the light was already full, as if it were closer to noon. But the sky was gray, with clouds hugging the hills north of town. The pilots told us we would have to wait. The planes, they said, didn't have any instruments—which is to say, no radar equipment. The pilots would be flying by sight, and if they couldn't see the mountains, well . . .

By 11:00 a.m. the sky had mostly cleared (emphasis on *mostly*), and we hopped into the aircraft. We would be flying 400 miles north in a pair of Helios Couriers, single-propeller planes celebrated by bush pilots for their reliability. I had no doubt that the forty-year-old machines were safe, but as DJ Spooky, Brune, and I clambered into one of them I got the feeling we would be traveling in the Volkswagen Bug of the sky: four seats packed into a space no bigger than a dining room table, our gear roped in behind us, every gauge analog.

We lifted off, made a turn over Fairbanks, headed northward. The planes passed over a line of silver zigzagging through the green woods—the Trans-Alaska Pipeline, pumping crude down from the Prudhoe Bay oil fields. For a while we could spot cabins and the occasional gravel road. And then there was nothing but the vast sweep of the bush. Hills after hills, dotted with the cones of spruce trees and countless nameless ponds. About ninety minutes north of Fairbanks we flew over the broad Yukon River and entered the mountains. That's when the importance of visibility became apparent. We wouldn't be flying over the mountains—we would be flying *through* them.

After the last spruce and the final fir faded away, the landscape emptied out even more. The mountains were stripped to their essentials: rock ribboned with water, and slopes of green that I assumed to be grasses and moss, and little else. The terrain was a work in progress. I remembered that the Arctic is one of the youngest ecosystems on Earth. Just yesterday (geologically speaking), the region was covered in thick sheets of ice, and so it has had little chance to develop. The deep freeze of the long winters slows down time further; it can take fifty years for a tree to grow to five feet. The empty valleys looked half-formed, as if Creation had been suspended partway through Day Three.

DJ Spooky and I were in the rear seats, and only Brune, sitting in the front, had a headset to talk to our pilot, Daniel, but it was clear we were having trouble. We headed into a mountain valley only to find the end enclosed by clouds. Suddenly Daniel made a sharp U-turn. A cliff face spun past the wingtip, the sharp peaks whirling just beyond our little cockpit.

We probed a second valley, but again hit a wall of clouds and had to make another nail-biting reversal. Incredibly, Daniel was navigating from memory. An Alaska Native, he had grown up in a cabin north of the Yukon River, and he knew the mountain range as well as I know the streets and alleys of San Francisco. But he couldn't find a way through. The passes were all closed. By this time, I had emptied the contents of my stomach into a couple of clear Ziploc bags and poor Paul Miller was squished against his side of the plane, trying to stay out of my spray.

After one more futile attempt to find a gap through the clouds, we made a beeline back southward, headed for Arctic Village, a hamlet situated at 68° North latitude that is home to an Alaska Native people called the Gwich'in. Stopping in Arctic Village hadn't been on our itinerary, but until the clouds cleared we would have no choice but to wait there.

The Gwich'in's traditional lands stretch from Fort McPherson in Canada's Northwest Territories across the US border to the east and north of the Yukon River, and Arctic Village (or *Vashraįį K'oo* in their language) is one of the nation's older settlements. A scant 200 people live there. The houses are mostly US-government issued prefab cottages painted a cheerful array of colors. "Downtown," as I heard one local refer to the crossroads at the center of the village, consists of a one-room grocery, a dilapidated Episcopalian church, the offices of the Gwich'in Steering Committee, and a sparkling new US Post Office and elementary school. Beyond that there's no sign of civilization, only a vastness of forest, lakes, and mountains.

After we landed at the gravel airstrip at the edge of the village, some of the Gwich'in got us set up in their community hall to wait out the rain. Our main host was Sarah James, a longtime member of the Gwich'in Steering Committee and one of the most vocal opponents of oil drilling in the refuge's coastal plan. Round-faced, with long, silver hair, James zipped around the village on her ATV to enlist her nephews in making that sure we had fresh water and firewood for the stove. (There is no water system in the town; villagers get their water from a communal pump house that draws from the Chandalar River.)

The Gwich'in are famous for being uncompromising about their sovereignty, and the tribe's spirit of resistance was on display there in the community hall. Protest banners hung from the walls of the log building. "Save the Earth," declared one that sported the iconic NASA photograph of our planet from space. "Save Gwich'in Way of Life," another read, and below that, "Our culture is not for sale. Support → Wilderness."

In 1971, Congress passed the Alaska Native Claims Settlement Act to deal with aboriginal land titles in the young state. Under the law, Alaska Natives (about 100,000 people at that time, in some 200-plus tribes) were asked to surrender some of their traditional lands, and in exchange they would receive cash payments as well as the opportunity to set up native-run corporations that would manage the mineral, timber, and petroleum resources in their areas. Many Alaska Native leaders supported the plan, and most tribes gave up their lands and formed corporations. The Gwich'in did not. Instead, they decided to hold onto their ancestral territories, a huge spread of land on the south side of the Brooks Range totaling 1.8 million acres, an area larger than Delaware.

The Gwich'in's commitment to the inviolability of their territory holds strong today. In the 1990s and early 2000s, Republicans in Washington pushed to open up part of the Arctic National Wildlife Refuge to oil drilling. Naturally, the Sierra Club and other conservation groups fought back. In that fight, the Gwich'in provided key moral leadership. They said the refuge should be given permanent wilderness protection, and they warned that drilling would disrupt or destroy the calving grounds of the caribou herds that tribal members still depend on for a significant amount of their protein.

I got that backstory from Ritzman and some Gwich'in who were hanging out at the general store as our group explored the town and whiled away the time. When the hours-long dusk began to settle over the valley, Sarah James invited us to her brother's house for moose stew. Walking around Arctic Village earlier, I had already taken notice of Gideon James's place, which seemed to be one of the few where people still kept a dog team. Harnesses hung outside his front door, and there were a half-dozen animals barking in his yard. (Gideon later told me he keeps eighteen huskies.) At seventy-five, he was still living some of the old ways. The front of his log cabin (no government prefab for him) was part machine shop for repairing chainsaws and snowmobiles and part hide-tanning workshop. Four rifles stood in a gun rack. As we entered, Gideon apologized for the ripe smell. He had shot a moose a few days before, and had been working on the hide indoors during the rain.

Rachel Maddow's cable news show played on a TV mounted in an upper corner of the long living room–kitchen. Along one wall Gideon had created a magazine-clipping shrine celebrating the life of Russell Means, a leader of the American Indian Movement in the 1970s. "Indian Country," proclaimed a banner with a red silhouette of a caribou on a field of turquoise. Sarah ladled the moose stew into bowls. Cooked with carrots and onions and served over macaroni, it tasted like the beef stew your mom might have made.

As happens among strangers, the talk turned to the weather. Gideon and Sarah agreed that all of the rain in the middle of summer was odd. "It's too warm, and it's too wet," Sarah said. "It didn't used to rain like this in the summer time." The rainy summers were just one of many changes in the weather that the brother and sister had witnessed. Gideon said, "It's a lot more warmer. It don't get that cold no more in the winter time. We used to get ice that thick"— and with his hands he measured out about four feet of length. "Now it's just barely *that* thick," and he cut the distance in half.

The changes wrought by global warming are happening right outside Gideon James's front door, and he expressed frustration that the question of whether humans are causing the problem remains a matter of political debate. "It's happening, though," Gideon said. "It's happening because the weather never cools off. It's dangerous. It's scary. Some people, they don't understand. Just like some of our state legislators don't believe it. Some of our leaders don't believe it. Some of these people who don't believe it, they got kids. They got grandkids. They need to think twice about it. But big money people like millionaires, they go and lobby them and they believe them."

Rachel Maddow carried on in the background. Gideon said, "This Earth is not balanced no more. It's like this," and he tilted his hands at a forty-five-degree angle.

Gideon James is right, of course. The science is unequivocal: the planet is warming, and human activities are to blame. Since the start of the Industrial Revolution and the beginning of fossil-fuel burning, atmospheric concentrations of carbon dioxide have increased from 280 parts per million to more than 400 ppm today, the highest level in about 650,000 years. From 1880 to 2012, average surface temperatures rose about 0.8° Celsius. But that's a global average. The Arctic has warmed disproportionately fast, between 2° and 4° Celsius. This is largely because of the historically unprecedented melting of Arctic sea ice, which has changed the region's albedo—that is, its reflectivity of the sun's heat and light. While white ice reflects heat, a dark ocean absorbs it. The steady diminishment of sea ice makes the region warmer, and in the process creates a feedback loop in which even more warming occurs.

Global warming is obvious in Alaska today. Permafrost is no longer so permanent, and as the ground melts, roads and power lines across the state are beginning to buckle and twist. A bark beetle infestation has hammered the state's conifer forests, since warmer winters and longer summers allow the beetles to reproduce more easily. As the forests die, wildfires (always a force on the landscape) have become more intense. Berries and trees are moving northward. Migratory birds like sandpipers and phalaropes are nesting up to a week earlier. Polar bears have moved inland and mated with grizzly bears, creating a new hybrid animal dubbed the "pizzly bear."

Meanwhile, coastal erosion and increased flooding threaten some Alaska Native communities. At least three native communities—the Inupiaq villages of Kivalina and Shishmaref, and the Yup'ik village of Newtok—likely will have to be abandoned and their residents relocated due to rising seas. The relocations are expected to cost tens of millions of dollars.

It's a patent injustice: the people who have done the least to cause climate change are the ones experiencing its impacts most harshly. Even the most remote regions are entangled with civilization, as the Gwich'in know all too well. "If you have polluted air, it doesn't go up into space and just go away. It stays here with us," Gideon said as we ate our moose stew. "When I worked for the tribe, I traveled down to the Lower 48 many times. I saw those goddamn eight-lane and sixteen-lane freeways going with cars twenty-four hours a day. All of that carbon dioxide, where does it go? It doesn't go nowhere. It stays right here with us."

He paused for a moment, embarrassed to be preaching, and said apologetically: "I get carried away talking about this stuff. I don't care if someone believes a different way. I don't care. But I know what I'm talking about."

Along with the sixth mass extinction, global warming is among the clearest signatures of the Anthropocene. We have frayed one of the most important Earth systems—the atmosphere—and in the process we've altered the oceans, too, as the seas absorb much of the excess carbon dioxide and thereby become more acidic. Global climate change, more than any other phenomenon, seems to make all of Earth into an artifact. Whereas nature once encircled civilization, now civilization (or at least our effluent) encircles nature. This change in our relative position is part of what makes the Anthropocene feel so disorienting.

For the Gwich'in—who remain at an inflection point between traditional ways and modernity—the disorientation of the Anthropocene is especially acute. Gideon and Sarah James are both gray-hairs, and would be considered elders by most of the people in their community. Yet when they spoke of "The Elders," it was evident they didn't mean themselves but rather the old folks of their own childhoods. It sounded to me like they were discussing a mythic people from a long-lost age, whose wisdom, essential though it is, may no longer be as useful as it once was. The baseline is shifting too fast.

"I know this husband and wife, they are out there in the bush all the time hunting and fishing, and they know the land and everything," Sarah said as we did the dishes. "But she went right into the river on a snowmachine when the ice was too thin. Because things are changing. The knowledge that we've had for thousands of years is somehow changing. We have to relearn what's going on in order to tell our people which way is safe."

The next day opened with spitting rain and overcast skies. Everyone was bummed. It seemed like we would never make it over the mountains. Then, in the afternoon, the pilots got a weather report from the north side of the range: the clouds were clearing. We said quick goodbyes to our Gwich'in hosts and hustled to the airstrip.

Once in the air, we again dodged our way among the valleys looking for a way through the fearsome peaks. Beneath the wings, summits split into talus slopes. Rocks in swirls of orange and black framed strange, iridescent ponds. The airplane chugged over one last pass, skirted an ice field, broke into clear skies. We could see the coastal plain stretching to the northern horizon, a seemingly endless expanse marked only by the oxbows and switchbacks of rivers and hundreds of water-filled potholes. From above it looked as if mirrors had been flung across the land.

Daniel brought us in for a landing on a dry shelf on the west bank of the Aichilik River. As the plane descended, the engine noise scared up a small herd of caribou that had been grazing near the makeshift airstrip. The animals charged in front of the plane as we made a touchdown on the tundra.

We unloaded the plane and, looking around, saw that the one group of ungulates wasn't alone. Caribou covered this broad valley in the foothills. There were thousands of them, in numbers beyond count, dots of beige sprinkled across the green slopes. The scene was like a polar Serengeti.

"Whoa," DJ Spooky said. "It's so primeval." Ten minutes later I saw him trying to approach a small clutch of the animals. He walked toward them slowly, deliberately, his palms extended out in a sign of peace, the gesture somehow Vulcan and Spock-like.

Soon enough the second plane came in, unloaded, and, just like the plane before it, quickly took off, leaving us in the middle of the wild. We circled together, and river guide Peter gave us a brief orientation. For four days we would be paddling about fifty miles northward through the coastal plain until we hit the Arctic Ocean and the shores of Beaufort Sea. Peter then delivered a quick safety briefing. Always tell someone where you're headed if you go out alone for a hike. Always take your bear spray with you. And—really important—remember that the bear spray only works if you're standing upwind.

We had no idea how quickly that last bit of advice would came in handy. We were finishing up a lunch of crackers and smoked salmon when we spotted a grizzly bear upriver. It was about a mile away, on the far side of the water, a tawny lump plodding its way along the riverbank. A stiff breeze was coming through a gap in the hills, meaning that the bear was downwind and likely aware of our presence. The bear then crossed the river, apparently determined to investigate what we were.

Our photographer, Micah Baird, wanted to get some shots, and he talked Ritzman into moving upstream with him. The next thing I knew, they were in retreat, hollering and shouting and firing their bear spray. Bursts of cayenne-orange mist swirled in the air, to little effect. Ritzman and Baird were still retreating.

The grizzly came to the edge of the small bluff above our camp site, no more than twenty feet away. It was beautiful: a blonde beast, its fur the color of ripening wheat, with a dark snout and dark paws and the classic ursine expression that suggests an especially smart dog. It was also enormous, probably 500 to 600 pounds, with claws as long and as lethal as buck knives. On a short walk I'd taken right before the bear appeared, I had seen the evidence of what it could do—places where the ground had been peeled away as a grizzly tore apart the topsoil in a hunt for ground squirrels. Huge chunks of sod as big as coffee tables had been tossed aside like so many pebbles.

The face-off was intense. We yelled and shouted, banged together pots and pans, but the bear was unimpressed. All of our hollering,

I'm sure, just sounded to it like a bunch of high-pitched squealing. It sat down on its haunches, looked at us curiously as we kept up our shouting. Then—in a priceless display of cuteness—it rolled over and began scratching its back against the ground. In unison, we all went *awww*. The bear popped up again—suddenly we sounded even less fierce than an instant before. The bear lumbered along the bluff, as if planning to come down and join us on the flat. That's when Peter decided to break out the 12-guage.

Fortunately, the shotgun wasn't necessary. We kept shouting and yelling, and, just as suddenly as it had approached, the grizzly turned around and fled. It loped up the hillside and away from our camp, probably more interested in hunting squirrels.

After the relief came delight. A grizzly bear almost in camp! Caribou by the thousands! We had been in the Arctic National Wildlife Refuge for less than an hour, and already the place had delivered.

Alaska—especially the Arctic National Wildlife Refuge—has long been an emblem of the wild, which of course is why I was so eager to go there. Brune had called Alaska "a bold-faced name of global wilderness," a place in the same pantheon with the Sahara and the Amazon. Or, as Brune's long-ago predecessor, John Muir, once said, "To the lover of pure wilderness, Alaska is one of the most wonderful countries in the world."

Alaska's brutal winters and sheer size are a formidable defense against civilization's encroachment. The state is as big as Texas, California, and Montana combined, yet is home to a scant 735,000 people, with almost half of those living in Anchorage. Most of the state remains roadless, accessible only by bush plane. Roughly 45 percent of Alaska is legally preserved in some form—either as national parks (the state has seventeen), wildlife refuges, or state or federal forests. Congress has designated 58 million acres there as wilderness, meaning that more than half of the United States' legal wilderness is in Alaska. To crib a line from Wilderness Society cofounder Bob Marshall, Alaska is "a permanent American frontier."

Sure, that idea of the frontier can seem historically blinkered, and it's hard not to smirk when you spot the state's license plate motto—"The Last Frontier"—on a huge rental RV wending its way through the tourist-clogged roads of the Kenai Peninsula. But, as TV channels full of Alaska-based "reality shows" attest, the place remains the last retreat for those who want to test their spirit against primitive conditions. The huge, unpeopled spaces and the unforgiving weather are an irresistible lure for wannabe trappers, modern-day prospectors, or just plain old misfits and hermits. In *Coming into the Country*, his classic exploration of the Alaskan bush, John McPhee describes people who went to Alaska and found there "a wildness that is nowhere else." In Alaska, McPhee wrote, the silence "can be as wide as the country."

Sometimes the thirst for adventure ends in disaster. Who can forget Chris McCandless, the fool or romantic hero (take your pick) who starved to death or accidentally poisoned himself (take your pick) while trying to live alone in the Alaska bush. Before McCandless was immortalized by a best-selling book and a Sean Penn–directed movie, a writer for the *Anchorage Daily News* summed up the takeaway from his death: "The Alaska wilderness is a good place to test yourself. The Alaska wilderness is a bad place to find yourself." Alaska is a land that will accommodate just about any self-mythologizing . . . until it leaves you dead.

Located in the far northeast corner of the state, the Arctic National Wildlife Refuge is like Alaska squared, a place that ranks among the purest distillations of the American wilderness ideal. The campaign waged in the 1950s to protect the region paralleled the broader effort to pass the Wilderness Act, and the arguments deployed for and against the Arctic refuge anticipated many of the claims made for the wilderness legislation. Opponents of the proposal to create an 8-million-acre preserve straddling the Brooks Range complained that it would "lock up" the land from future development, especially mining and oil extraction. Locking up land, conservationists said, was *exactly* the point. Preservation of the Far North transcended the protection of mere scenery for the purposes of human recreation and aspired to a higher goal: landscape-scale conservation in order to sustain biological processes.

A pair of biologists, Margaret and Olaus Murie (then head of The Wilderness Society), spearheaded the campaign to protect the forested mountains and the open tundra north of the Yukon River. As a field researcher with the US Biological Survey (now the US

Fish and Wildlife Service), Olaus had done groundbreaking work in wildlife biology. Margaret—or Mardy, as everyone called her—was also an impressive naturalist, the first woman to earn a degree from the University of Alaska. For their honeymoon the couple canoed and dogsledded 550 miles around the upper reaches of the Koyukuk River conducting caribou research, an experience chronicled in Mardy's book *Two in the Far North*. Amid the valleys of the Brooks Range, the couple found, in Olaus' words, "a place to contemplate and try to understand our place in the world."

In May 1956, the Muries and three other biologists flew to the headwaters of the Sheenjek River about 100 miles northeast of Arctic Village to spend the summer conducting research. They camped at a spot they christened "Last Lake," where it was not unusual for thousands of caribou to cross the river. The herds created a sound that reminded Mardy of "a freight train roaring through the valley." The group was later joined by Supreme Court Justice William O. Douglas and his (then) wife Mercedes, whom the Muries had recruited to boost their political efforts to protect the region. The months-long experience crystallized the Muries' thinking about the aesthetic and biological importance of the area and gave them the intellectual ammunition for the political fight to come.

During the Congressional battle to get lasting protection for northeast Alaska, conservationists deployed the classic argument for wilderness as a civic resource. Channeling Teddy Roosevelt, Olaus said this "last great wilderness" could provide "that precious frontier atmosphere which helps build a strong civilization." But the area was also the icon for a more profound conservation goal: it was one of the yet-unbroken landscapes where life's ancient rhythms could still be witnessed.

Nothing symbolized this more than the huge caribou herds that freely roam the region, a spectacle that recalls the vast populations of bison that once dominated the North American Plains. In the course of a year, the 150,000-animal-strong Porcupine caribou herd walks at least 1,000 miles. The herd moves in a steady, counterclockwise circuit from its calving grounds on the coastal plain, through the mountain valleys of the Brooks Range, to its wintering grounds in the Canadian Yukon, and back again, following available forage and ancient instinct. Phenomena like that, the biologists believed, could inspire people to the ideal that in some places wild nature should maintain its primacy. "The environment is not tailored to man," Mardy wrote, "it is itself, for itself."

The expedition participants returned from Last Lake determined to create a wilderness preserve on an unprecedented scale. Here was a chance to do ecosystem preservation right. "This is—and must forever remain—a roadless, primitive area where all food chains are unbroken, where the ancient ecological balance provided by nature is maintained," Justice Douglas wrote.

Heartfelt though they were, such arguments failed to convince legislators in Washington. The bill to protect northeast Alaska died in Congress in the fall of 1959. But less than a year later President Eisenhower's secretary of the interior, Fred Seaton, signed an order to make the region a federally protected wildlife range. In 1980, the range became a permanent refuge and was expanded to its current size of 19 million acres when Congress passed and President Jimmy Carter signed the Alaska National Interest Lands Conservation Act.

Yet the place remains contested terrain. During our pilgrimage into this vast wilderness we would be threading our way between two distinct worldviews. On the eastern shore of the Aichilik River lays the Mollie Beattie Wilderness, an 8-million-acre expanse that is the second-biggest legal wilderness in the United States (named after the first female director of the US Fish and Wildlife Service). On the west shore of the Aichilik River is what, in government parlance, is known as the refuge's "1002 area" (or "Ten-O-Two," as everyone calls it.) Under the 1980 legislation, this 1.5-million-acre section of the coastal plain was set aside as a "study area" for possible petroleum production. When you hear people talk about oil drilling in "An-Whar," the Ten-O-Two is what they mean.

(In January 2015, six months after our rafting trip, President Obama directed the USFWS to begin managing the entire refuge, including the Ten-O-Two, as wilderness, which would prohibit oil drilling. But official wilderness designation requires an act of Congress, and, given the attitudes of this current Congress, that seems a long shot. The refuge is still disputed ground.)

To use, or not to use? That is the question posed by wilderness. The oil and gas companies say "Use it." Environmentalists and their Indigenous allies say "*No.*" Sarah James told me that before the arrival of the whites, the Gwich'in had no word for "wilderness." Indeed, this is the case for most hunter-gatherer societies, for whom there is no sharp dividing line between the domesticated and the wild. During an unprecedented 1988 summit of all the tribe's members to discuss the fate of the refuge, the elders took time out to come up with a Gwich'in expression synonymous with the word *wilderness.* "It took them a long time," James said. They finally settled on the phrase, *The place where life begins.*

"Leave it the way the Creator made it," she told me. "That way we will know it is protected."

If only it were still that simple. In the campaign for the establishment of the refuge, Olaus Murie wrote that the question of northeast Alaska "involves the real problem of what the human species is to do with this Earth." It's a problem that, as the days on the river would reveal, has been sharpened by the arrival of the Anthropocene. The question is not just to drill or not to drill, but whether any place can remain as "the Creator made it" and still be "protected."

❧

The day after the grizzly bear encounter was our first on the river, and it took us a good couple of hours to break camp and learn how to load the two rubber rafts. By the time we were ready to go, our watches said it was well past noon. The late start was also due to the fact that a few of us had already taken on the sleep schedule of Berlin ravers. We'd stay up well past midnight, to when the light was dreamiest, then sleep until late morning. Not that we were worried about time. The sun hadn't set at all the "night" before, and it wasn't as if we were going to run out of daylight. Still, due to the unplanned layover in Arctic Village we were a day behind schedule and needed to make up some river miles. It felt good to shove off into the chocolate-milk-colored waters of the Aichilik and point the rafts northward.

Much of the time we were able to just go with the flow. Now and then Ritzman and Peter, the boat captains, would call out a command, and we'd strike our paddles into the water to keep our distance from the sand shoals and gravel bars, or to maneuver through the quick waters at the cutbanks. Our trip had been timed to hit the tundra's two-week-long spring, the narrow window between ice-melt and the explosion of voracious mosquitos and black flies. But even in the week of the summer solstice, ice clung to the riverbank in many spots. Huge blocks of ice were stacked along the shore, slabs of pure white and digital blue that made me think of a deep-freeze layer cake. The day was warm—probably in the low seventies—and the ice was melting fast. As we paddled downstream the rafts zipped between colonnades of waterfalls, the meltwater torrents splashing around us.

Other than the river murmur, the scene was quiet. A few times we scared small flocks of mergansers up out of the river, and once we startled a ptarmigan from the riverside willows. In the mid-afternoon we spotted a caribou herd cooling itself on a far-off ice sheet. Even from a distance, we could pick out the signature sweep of their antlers (which both bulls and cows sport), the U-shape reminiscent of a football goalpost. But we saw nothing like the herds of biblical proportions from the day before. The land was empty and still.

We had been in the great wilderness for only a day, and already the place had disoriented me. At the top of the world at the height of the summer, direction had become all but meaningless as the sun refused to set. The never-ending light had also blown apart my sense of time, turning a.m. and p.m. into abstractions. My grasp of distance had dissolved, too. Without any trees to speak of, it was impossible to get perspective, to tell if something was half a mile or three miles away. The landscape was oceanic; if I looked hard enough at the distance it seemed like I could see the girth of the globe. It was the kind of wilderness that causes a psychic vertigo.

As we beached the rafts and set up camp, a few of us tried to make sense of what we had experienced so far. "It's just beautiful. It's like a great painting," said Shirley, an avid watercolor painter. "It does something chemical to the brain. The space—it's so huge, it makes you feel inconsequential. You realize that we're not in control."

Rue had a similar take: "Out here, our will doesn't mean that much. Think of that grizzly bear. We're not at the top of the food chain. Stripping away all of the noise of our daily lives"—no doubt a welcome relief for a woman who has three kids at home—"you have to face yourself, you have to set aside your ego. Think about Jesus, and John the Baptist. They went into the wilderness to get clarity of purpose."

Me, I needed more time to think and to be by myself. As the rest of the group started making dinner, I set out to hike up a small hill that looked to be about a mile away (turned out to be more like two).

From a distance, the tundra appears flat and even, but up close it's a different story. I soon found that the land was split by depressions and small humps, the result of millennia of frost heaving and summer melt. Technically, the tundra is a desert; in an average year the North Slope receives just seven inches of precipitation. But because of the permafrost there's nowhere for the water to drain, and so precipitation collects in the first couple of inches of soil, making the tundra a weird kind of swamp.

The plain, I quickly discovered, was dotted with sedge- and rush-filled bogs that appeared without warning. In other areas, the turf was propped into fields of tussocks—foot-wide mini-mounds rising up like mushroom caps. To walk cross-country meant either slaloming among the tussocks (a squishy affair that quickly resulted in soaked boots) or else trying to bound from one to the other (which felt like being dropped into the original Super Mario Brothers video game). Bouncing from tussock to tussock was like walking on a spring-loaded carpet. No matter which approach I tried, navigating the terrain was exhausting.

I hadn't gone far when I realized that (as Barry Lopez points out in his book *Arctic Dreams*) on the tundra you're walking *on top of* a forest. I bent down to inspect. On each tussock there were as many as four or five varieties of *salix*—that is, willow. But the willows were no more than three inches tall. Alongside these dwarf "trees" were several different types of grasses, and among the willows and grasses grew a lilliputian universe of moss and lichen. The mosses came in a psychedelic rainbow of colors: dusky sage, forest green, glow-stick green, rust, orange, pink.

The flowers were also miniature. There were patches of tiny pink moss campion, delicate purple cress, and the small white bulbs of moss heather. Among a clutch of mini-willows I found pink shooting stars just like the ones we have in California, only the bloom was shrunk to a quarter of the size. The elven-scale world underfoot felt incongruous with the gigantism of the open plain and the epic mountains, and as I watched the dances of the small Lapland longspurs (a bird practiced at cuckoldry), I had to laugh: such tiny dramas, enacted against a heroic backdrop.

I was marveling at the scene when I spotted a single caribou across a stream draw about a half-mile away. As I reached for my binoculars, a flash of white darted into view. The caribou easily outran its aggressor, and there was no chase. I glassed the hillside and saw it was a red fox that had caused the caribou to run.

It was by far the biggest fox I had ever seen—from a distance it looked to be the size of a California coyote. Cinnamon red, with a large, bushy tail tipped white. I watched as it made figure-eights across the slope, hunting for the nest of a gull that was dive-bombing it. Frustrated in its egg search, the fox headed southward. I followed the fox with my binoculars as it trotted through a swath of tundra where a large herd of caribou had passed not long ago (leaving the ground trampled and grazed to a softer shade of green) and then crossed over a hill and out of sight.

When I got back to camp, I was excited to report my fox sighting. Ritzman and Peter, a couple of longtime Alaska hands, weren't all that impressed. But they were somewhat surprised. It was unusual, they said, to see a red fox this far north. Perhaps, they told me, it had something to do with climate change. Recent studies had revealed that as the weather warms, the red fox's range had expanded farther and farther onto the tundra.

The red fox's larger range has been a detriment to its smaller cousin, the Arctic fox. Red foxes had been seen digging into the dens of Arctic foxes to kill the kits and sometimes even the adults. The news brought to mind a line from a report by the International Panel on Climate Change: due to global warming, "entire ecosystems will be forced to move, colliding with each other."

What had seemed a wonderful wildlife sighting took on a darker cast, making me think that what I had seen was something—I guess you could say—*unnatural.* Due to human activities thousands of miles away, the Arctic fox was becoming a casualty of climate change, much like the polar bear. But instead of mating with its nemesis to create the vulpine version of a pizzly bear, the Arctic fox was ending up as supper.

What could be done? What *should* be done? Was it time, perhaps, for humans to intervene, for state or federal wildlife officials to begin hunting the red foxes in order to even out the contest?

That might sound absurd—hunting one species of fox to save another. Yet that's exactly the kind of soul-searching dilemma conservationists face in the Anthropocene. The Human Age has thrown us into a *terra incognita* that we are having to learn to navigate on the fly. Increasingly, we don't just conserve ecosystems—we *curate* them. In the process we are forced to make life-and-death decisions about what we hope to protect and preserve, what we're willing to stave off and suppress.

Across the United States, federal and state agencies and large landholding NGOs like The Nature Conservancy are busy tinkering with the environment in order to save it. Faced with the ever-growing evidence of how we've busted many ecosystems, officials are confronting competing choices about what to value most. Should we prioritize preserving biodiversity, the bedrock of healthy ecosystems? Should we instead try to conserve a quality of "naturalness" in the landscape? Or, especially in designated wilderness areas, should we try to guard the condition of "wildness"?

These dilemmas are being fueled not just by the scale of human impacts on the environment, but by their speed. Ecosystems change; there's no such thing as a static landscape. But the changes now under way are occurring too quickly for many plants and animals to keep up. Global warming is shifting habitat zones toward the poles at the rate of up to four miles per year. Combine that with the fact that there are fewer places for plants and animals to relocate to (because the human footprint has overtaken so much habitat), and you can see how conservation biologists have found themselves in a chronic state of emergency.

In response, some conservationists have fallen back on a kind of biological "Pottery Barn Rule": We broke it, and now we own it, and therefore we have a responsibility to fix it. In the foreword to a 2010 book about the future of the national parks, National Park Director Jon Jarvis wrote, "As climate change is, at least in part, anthropogenic, the paradigm of allowing nature to rule the parks is no longer viable."

We now routinely supervise ecosystems that are otherwise undeveloped. On many a weekend, volunteers attack cheat grass in the Mountain West, target deep-rooted fennel on California islands, rip out ice plant along the Pacific Ocean shorelines. In the Northern Rockies, government agencies and private groups have planted hundreds of acres of white-bark pine saplings to counteract tree loss associated with climate change; the trees were specially bred to be resistant to drought and blister rust. To try to beat back invasive and water-sucking tamarisk in the canyonlands of the Southwest, officials have used chainsaws, artificial fires, and even released a pest, the tamarisk beetle, into the wild. On the Columbia River, government officials dedicated to protecting salmon and steelhead populations haze sea lions with flares and rubber bullets—and sometimes even shoot and kill the animals—to keep them from eating too many fish. In the forests of Oregon, federal wildlife officers stalk and kill barred owls to prevent them from outcompeting the smaller, and more endangered, spotted owl, a campaign that is nothing less than a conservationists' Sophie's Choice.

Sometimes these interventions occur in designated wilderness areas. In the remote Frank Church–River of No Return Wilderness of Idaho, US Forest Service workers spray herbicide in an effort to control spotted knapweed and rush skeleton weed introduced on the hooves of cattle and sheep decades ago. National Park Service employees have resorted to bulldozers to try to beat down Brazilian pepper trees in the heart of the Everglades. In an attempt to repair eroding soils, teams equipped with chainsaws have hacked away at juniper and pinyon trees in New Mexico's Bandelier Wilderness and mulched the ground with their sawed off branches. When U.S. Forest Service officials in the Saint Mary's Wilderness of Virginia realized that a river there had become too acidic, reducing the diversity of aquatic life, they used a helicopter to dump 140 tons of limestone to buffer the pH of the river. It worked. Macroinvertebrate diversity and fish density increased, though the treatment had to be repeated six years later.

Even more heavy-handed interventions are under consideration. If California were to enter into a cycle of mega-drought, park rangers might have to begin irrigating the stands of giant sequoia trees in Sequoia National Park. Rising temperatures could force government agencies to begin the assisted migrations of the Joshua tree or the pika, an alpine rodent that lives at the top of the mountains, and so has nowhere else to go as the planet warms.

These interventions and many others (the above litany is a short list) raise some tough questions. For starters, there's the issue of time-scale and commitment. Will we have to undertake such actions in perpetuity? There are also concerns about efficacy. Won't this be like the biological version of painting the Golden Gate Bridge—as soon as you finish, you have to begin all over again, only this time in conditions that likely will have changed since the task was started?

Then there are the worries about unintended consequences. What if our well-intended meddling goes awry? History is littered with such examples. Just think of the introduction of now-invasive kudzu vine, originally intended as cattle forage. Or this: in the late seventies, California officials poisoned lakes in the Sierras to beat back invasive fish and create a blank slate for the reintroduction of native golden trout. DNA tests later revealed that the introduced trout were hybrids. The lake poisonings likely killed the last true native trout.

Much like the community division over the oyster farm in Point Reyes, the question over whether to intervene in ecosystems or to leave them alone has caused an emotional rift among people with similar sympathies. In a book with the provocative title *Beyond Naturalness*, academics David N. Cole and Laurie Yung lay out the dilemma: "It is increasingly clear that just leaving nature alone will not be adequate to conserve biodiversity and many of the other values we associate with protected areas." Yet they wondered, "Does more human intervention make things better or worse? Can two wrongs make a right?" Christopher Solomon, a science writer, has argued in the *New York Times* that meddling with the wilderness might be a "necessary apostasy" at this point in time.

The day before we flew out of Fairbanks, I had the chance to talk through some of these issues with Roger Kaye, a US Fish and Wildlife Service staffer who is the wilderness specialist for the Arctic Refuge. Kaye, a mustachioed guy who has spent thirty years as a bush pilot ferrying biologists in and out of the refuge, is an opinionated fellow and an uncompromising wilderness advocate. (He was careful to let me know that his views don't necessarily represent the views of his agency.) Kaye believes that, at least in wilderness areas, we shouldn't intervene in biological process. Wildness should hold sway.

"I think wildness is the most objective reality in the cosmos," Kaye told me, "if you take the definition that I use: wildness is the condition of a landscape that is free from human intent to manipulate, shape, or control. Wildness is an evolutionary process. That is how it's been since the planet was born. How do you operationalize it? Easy. You put lines around a place and say, 'We will not apply our intent beyond this boundary. We will allow this area to evolve according to its own processes.' Biologists would say, 'It's not about intent; it's about effect on the ground.' They're trying to maintain a status quo, and there's a presumptiveness, an arrogance that we can maintain certain conditions that we like."

And what if that hands-off approach leads to the diminishment, the extinction even, of some species? "You just leave nature alone," Kaye said. "You don't do anything. Whatever happens, happens."

Not surprisingly, that's a difficult idea for many conservation biologists to accept. During interviews that occurred before and after my trip to the Arctic, some stalwart conservationists expressed concern about the possibility of sacrificing species for the sake of an ideal.

One of those skeptics is Reed Noss, an esteemed biologist who has spent much of his life studying, and trying to preserve, wildlife habitat. "I love those areas," Noss said of wildlands. "It's an emotional thing, a spiritual thing, even though I don't believe in spirits." But, he said, "the bottom line for me is that we have to stop the loss of biodiversity." Noss was cautious about making sweeping generalizations. He said he would want to consider the issue case by case and acknowledged that, for the most part, he just wasn't sure. "It's a philosophical challenge that I haven't completely settled for myself," he said. The main problem, he pointed out, is that there are relatively few wild places that are big enough and remote enough not to require *some* level of human manipulation. "Wilderness and biodiversity are my most cherished ideals, but in cases of conflict between the two, I would pick biodiversity, because we are in an extinction crisis. I would argue for intervention."

Another eminent ecologist, Duke University biologist Stewart Pimm, was more unequivocal. "Now we're hearing, 'We shouldn't protect species, we should be protecting wildness,'" Pimm told me, his English accent displaying no small amount of annoyance. "I am happy to discuss the idea in a pub over a beer. But other than that, it's a bad idea. It's not that I don't understand wilderness. It's not that I don't appreciate wilderness. If you talk about ANWR, I know what you mean by ANWR. I know what the birds are; I know what the animals are. I can count how many caribou. I can count how many Arctic foxes. I can count how many gulls. Those things I can measure. But if you are managing for wildness, how are you going to measure it? Where is your 'wildness meter'?"

To Roger Kaye, that's the kind of reductionist thinking that gets in the way. "Biologists!" he huffed at one point in our discussion. "Their whole paradigm is, if you can't weigh it and measure it, you can't explain it, and therefore it doesn't exist."

When it comes to the question of whether to intervene in wilderness, there seems to be—to borrow a phrase from the International Panel on Climate Change—a collision of *ideological* ecosystems. It's a clash of priorities and values among people of shared beliefs.

Peter Landres has spent a long time thinking about this question. He works at the Aldo Leopold Center Wilderness Research Institute, an interagency science think tank in Missoula, Montana, administered by the Forest Service, and has spent the last fifteen years trying to square the circle between the biological values and aesthetic-ethical values of wilderness. "This is really hard stuff," Landres, a PhD research ecologist and a self-described "generalist," told me. "I don't like to use the words *wild* or *wildness*, because different people use these in different ways. When some people use *wild*, they mean 'native species' and call this 'pristine.' But there is no such thing as pristine, which comes from a biblical orientation. So throw *pristine* out. Other people use it to mean something like 'self-willed,' and that's mostly about letting nature be by not interfering or manipulating it. So *wild* and *wildness* have become fuzzy, making it harder for people to talk about it. When talking about wilderness, the word I use is *untrammeled*. Untrammeled is the absence of intentional human manipulation, hindrance, or control—like, for example, not suppressing naturally caused fire."

Landres has spearheaded an interagency effort among the Forest Service, National Park Service, Bureau of Land Management, and US Fish and Wildlife Service to come up with a working definition of "wilderness character" and a way to monitor how it changes over time. The document—titled, appropriately, "Keeping It Wild"—will give federal officials on the ground just the kind of "wildness meter" Pimm says is missing. Still, Landres (who, for the record, leans strongly to the side of nonintervention in wilderness) acknowledges that even when federal-agency wilderness stewards have a firmer definition of wildness, they'll still have to decide *whether* or *how* to manage.

Landres said to me: "This is the crucial distinction—to separate 'natural' from 'untrammelled.' Is there anything that's still wild? Yes. Are ecological systems impacted worldwide? Yes. There's no question about that. When you do lake sediment or glacier coring, you can see the presence of the industrial age. You can see soot. You can see particulates. There is a pervasive [human] presence that has diminished the natural quality. But that does not diminish in any way the choices that we have about how we manage these systems. And that's why the untrammeled idea is so important, and why I used the title 'Let It Be' in my chapter in the *Beyond Naturalness* book."

For Landres, the choice of whether or not to intervene in a landscape scale isn't just a matter of biological science—it's also an ethical decision. At the end of our talk, he said, "Ecologically, our systems have already been compromised. In wilderness, the goal is not to have an ecological target. The goal is to allow evolution to be unfettered by the human desires, whims, drives, and nuances of the day—'the desires du jour.' That's the goal. Our ecological systems will continue as they have and they will, as long as we don't meddle with them. Are there reasons to meddle? In some cases, yes. The ethical importance of treating wilderness ecosystems with humility and restraint is of paramount importance right now in our vastly developed and anthropocized world. There are even *more* reasons to leave wilderness areas untrammeled now, in comparison to before."

❦

"The Greeks named the Arctic. It was *Arktos*, the constellation of the Bear of the North. So here we are, in another culture's imagination. And we have to push people beyond the frontiers of their own imaginations. We have to translate this for people who can't think beyond the city, who aren't even able to see the night sky to name anything."

That's a little sample of what it was like being on the river with DJ Spooky, who actually talks like that in casual conversation. Paul Miller is nothing if not an urbane cat—raised in a Washington, DC, townhouse that served as a seventies-era salon for the capital's African American intelligentsia, and now a longtime fixture of the Chelsea art and music scene. When he muses, which he does easily and often, Miller speaks in an erudite stream of consciousness that, in the space of a minute, can mix together Paul Robeson and John Cage, Jorge Luis Borges and Aristotle. It was now Day Three on the river, and I was paddling in a raft with Miller and Rue. We had just shoved off when Miller began waxing philosophic.

"Joyce began *Finnegans Wake* with the word *riverrun*," he said. "And here we are on a river. I love this: river as metaphor. Like this river, there are multiple paths that we can take into the future. Streams appear, they disappear. The lines fold in and out of one

another. You make the choices—to go left, to go right—and that
determines your fate and your future. Maybe you go forward. Maybe
you get trapped on a gravel bar. And if you get stuck, you have to
bounce. You spin in circles."

Metaphor indeed. Here I was spinning in circles, stuck trying to
puzzle out what conservation would look like in the twenty-first
century.

Oddly enough, advances in the science of ecology had complicated the issue. The activists and lawmakers who wrote the Wilderness Act and the Endangered Species Act focused on the goal of
keeping ecosystems in a "natural" state, a prescription that followed
the best knowledge of the time. In the mid-twentieth century, biologists believed in something called "climax ecosystems." A natural system—a forest, say—would evolve through a succession of stages until
it reached a state of dynamic equilibrium, a climax. This thinking
influenced the "Yellowstone Model" of conservation: draw protective lines on a map, and let everything inside remain as it had always
been. The notion of climax ecosystems was the bedrock of national
park management for close to half a century. In 1963, a National
Park Service document called the "Leopold Report" (written by
Aldo's son) recommended that "the biotic associations within each
park be maintained, or where necessary recreated, as nearly as possible in the condition that prevailed when the area was first visited by
the white man. A national park should present a vignette of primitive
America." The Wilderness Act decreed that a wilderness area should
be "managed so as to preserve its natural conditions."

Today ecologists know different. Biologists now agree that ecosystems are characterized not by equilibrium, but by disturbance. A
massive fire, a drought or some other change in climate can abruptly
shove an ecosystem in a whole new direction. Ecosystems are, above
all, "stochastic"—random. The new science confirms one of the
original insights of ecology: the only constant is change. But it also
complicates the day-to-day, on-the-ground work of conservation.
How do you preserve something in its "natural condition" if you
can't determine what "natural" is, if "natural" is ever-changing?

Nature has always been a squirrely idea, every bit as riven by contradictions as the concept of wilderness. I think science writer Alan

Burdick nailed it in his book about exotic species, *Out of Eden*, when he wrote, "Insofar as a single line can be said to separate the world of nature from the world of man, that line is exactly the thickness of the human skull." Now, in the Anthropocene, that line has gone squiggly.

Nature—as a word meaning something beyond humans—has lost its usefulness. This has been true at least since global climate change made itself felt. In the first popular book about global warming, Bill McKibben declared that by changing the atmosphere we had caused "the end of nature." By "nature" McKibben meant something apart from human civilization, which is the common, colloquial understanding of the word. But the usage elided the fact that we humans are also part of nature. The English word *nature* comes from the Greek, *natura*, meaning "birth." We are born from this Earth, and so we—along with our culture and creations—are also, essentially, natural.

To an ant, the anthill is an artifact, the biologist E. O. Wilson has observed. The quip illuminates the fact that making neat divisions between the natural and the artificial is never simple. Even Tokyo and Manhattan are, in a way, natural. Or take an easier example—the garden. A rose bush is sculpted by and for people. An apple tree is pruned every year, its branches shaped according to human needs. The rose-flanked lawn and the orchard are undoubtedly natural in their functions: they are living, breathing beings. At the same time, they are artificial in their form: they have been manipulated by our green thumb.

The classic definition of nature as a thing opposite to humanity is also problematic because it sets up the unhelpful idea of nature as "untouched." If we humans are outside of nature, then any intervention or incursion we make into nature is automatically a desecration. There can be only the pristine or the profane. Unfortunately, the persistence of the pristine-profane dichotomy has fueled the intellectual attacks on wilderness. Wilderness is impossible, the "neo-environmentalists" argue, because there is no such thing as a pristine place untouched by human actions. According to the authors of "Conservation in the Anthropocene," the environmental movement's "intense nostalgia for wilderness and a past of pristine nature . . . is both anachronistic and counterproductive."

But this is a straw-man argument. As Peter Landres pointed out, "wilderness" isn't synonymous with "pristine." The vast majority of environmental professionals—activists, scientists, and land stewards—have long understood that there is no pristine anywhere. "In fact," writes Michael Soulé, one of founders of the science of conservation biology, "educated conservationists have not believed in the existence of pristine places or systems since at least the 1970s, when DDT was found in animal tissues everywhere, including in the milk of human mothers." In one of our conversations, Dan Ritzman was even more blunt, as if the point were obvious: "No place is unmarred anymore, right?"

The advent of the Anthropocene makes the definition of *natural* even more confounding. We cast our shadow everywhere, making it impossible to pinpoint the border between the "natural" and the "unnatural." Perhaps the clearest proof of the uselessness of the word *nature* is the fact that it no longer works without a modifier. Peek at the academic literature and you'll find that the preferred terms of art are "nonhuman nature," or "more-than-human nature," or, simply, "wild nature." When a word can't stand alone, it has become unmoored from meaning.

If "nature" is an anachronism, it would seem to doom the conservationist endeavor. Unless, that is, we find another ideal on which to set our aspirations for protecting the last undeveloped parts of Earth.

As we paddled northward to the Arctic Ocean, it seemed to me that the best ideal available to us is wildness. Forget untouched. What matters now is whether a place is *uncontrolled*. On a post-pristine planet in which "nature" no longer makes sense, protecting the wild is more important than ever.

It's a big ask—to suggest that what we need now is to give up control. After all, humans are tinkerers by, um, nature. Our instinct is to manipulate, and so to ask that people let go of our dominance is difficult.

At least, that was Brune's view. Surrendering control in an era of uncertainty? It seemed to him a nonstarter. "What's the meaning of wilderness in a time of great change?" he said later that day, after we had made camp and eaten supper. "If wilderness remains key in the twenty-first century, it's going to be more of a grounding thing than a letting go. It's going to be about finding yourself rather than losing yourself."

Brune and I were on a post-dinner constitutional across the tundra. By now our group had traveled far out on the coastal plain, and the mountains had been reduced to a shadow in the south. The light had mellowed to the soft yellow of eternal evening. What had to be millions of white avens and buttercup anemones and alp lilies stretched beyond sight. The whirls of wildflowers reminded me of constellations. Given such a show, who needs the stars?

Brune is a thoughtful guy. When asked a question, he'll take his time to work out the answer, and if he doesn't have one, he's not afraid to say, "I don't know." I had asked him where, exactly, wilderness and wildness fit in civilization's twenty-first century survival kit. Even for the head of the Sierra Club, the answer was not as self-evident as it would have been a generation earlier.

"I'm still trying to figure it out," he admitted. "I do know, at least, that we need these refugia as places we can recover, physically and mentally. I know that we have to connect people to place. To encourage a love of place, or at least a curiosity of place. And then, hopefully, they'll find a connection with nature, and they'll go do something."

We walked some more, tracing a big loop back toward our riverside camp. A thick fog began to approach from the ocean, now just a day's paddle to the north. I asked what he thought about manipulating the wild in order to save nature. Brune paused, then looked at me and said, "I don't know. It seems like we need a recalibration. We have to find a way to be comfortable with ambiguity."

True enough. When President Lyndon Johnson signed the Wilderness Act, he declared that the law would "preserve for our posterity, for all time to come . . . this vast continent in [its] original and unchanging beauty." Today we know what an impossible aspiration that is. Wilderness can no longer be the kind of anchor it once was, a promise of constancy and a reassurance in a changing world. Even the wildest, least-developed places will shape-shift. And as wildlands change, the changes will confirm how little we know.

Perhaps that reminder of the limits of our understanding is itself one of the key lessons of wilderness in this fraught new age. Again I thought of Thoreau, who suggested that the wild can move us toward a kind of epistemological humility. Wilderness can force on us a Socratic wisdom that our knowledge is dwarfed by how much we don't know. "The highest that we can attain is not Knowledge," Thoreau wrote in his essay "Walking," "but Sympathy with Intelligence."

Be comfortable with ambiguity. With his admission of uncertainty, Brune had gotten me a little closer to the certainty I desired. It's precisely because we understand so little about how our interventions might affect ecosystems that we should act with a surplus of caution. The very ambiguities of wildness in the Anthropocene make a case for taking the more hands-off approach.

In our Fairbanks chat, Roger Kaye had reminded me of something that Howard Zanhiser, the chief author of the Wilderness Act, once said: "The essential quality of the wilderness is its wildness." That sounds merely tautological, vacuous even, until you remember that wildness means self-willed. Wilderness is not about purity or even primitiveness. It's about autonomy—letting things go their own way, even when we're convinced that we know what's best.

We were on our last day of river paddling. As the Aichilik approached the ocean, the river split into a dozen different braids, and we kept getting stuck on the gravel shoals, forcing us to climb out of the rafts, slog through the shallows, and then push off again into the mainstream. In some places, huge chunks of the riverbank had collapsed into the water, leaving the earth turned sideways, the wildflowers growing perpendicular to the sky. The caribou had disappeared, but now the air was full of birds. We saw many pairs of jaegers, long-tailed and sharp-winged, like gulls with style.

As we got closer to the Arctic Ocean the temperature dropped, and we had to haul off the water so everyone could put on an extra layer or two. In front of us loomed what looked to be a giant wall of ice a hundred feet tall stretching across the horizon. (In fact, when we reached the ocean we would find only scattered ice floes; what we were seeing was an optical illusion called a *fata morgana* that distorts the size of objects on an empty plain.)

At the river's mouth we stopped to inspect the ruins of an Inupiaq village. Thick beams of wood were melting into the turf. Flowers covered the remains of an iron stove, the line between nature and culture dissolving once more. An Arctic owl—giant, snow-white—watched us impassively.

Then the final push: battling the currents, we paddled across a broad lagoon until we arrived at a spot marked on the map as "Icy Reef." It was a long, thin stretch of sand and gravel wedged between the lagoon and the sea. The ocean was perfectly calm, the water gunmetal blue and dotted with small icebergs that had half-melted into weird, twisted shapes. Driftwood covered the strand, great piles of bone-white timber that had sloughed off the forests of Canada and Alaska and washed to the sea. A pair of terns guarded this strange kingdom, and they dive-bombed whenever anyone got too close to their nest.

Looking down the beach, I spotted something out of place—a bright blue square about a hundred yards away. Shirley and I walked down the sand to inspect. As we got closer and realized what we were seeing, shock walloped us.

It was a large plastic cooler on wheels, with a white plastic handle for towing and the words *Polar Roller* emblazoned on the side. Foam insulation was coming out of the inside, and the cooler was covered in pockmarks, as if someone had attacked it with an icepick. More likely, a polar bear had used the Polar Roller as a chew toy.

And so it has come to this: even at the ends of Earth, you find yourself reduced to a beach cleanup.

The cooler—so incongruous, and such an offensive sight in that place—confirmed for me the necessity of keeping a few places free from human manipulation. If we know that we will inevitably mar Earth with our accidents, the very least we can do is keep some

places free from our intentions. If the wild is going to remain meaningful, we'll have to commit to leaving our hands off, no matter the
consequences.

I want to make sure I am perfectly clear: Most of the world
will continue to be domesticated, a human realm of cities, suburbs,
farms, reservoirs, timber plantations, and aquaculture pens. What we
are talking about here is a sliver of the planet, the last pieces of
the untrammeled Earth. In the United States, remember, designated
wilderness constitutes just 5 percent of the national territory. Is it
really too much to ask that those wildlands remain outside our overt
dominion?

By keeping the wilderness autonomous, we'll make sure that
we still have spaces that can serve as a "base datum of normality,
a picture of how healthy land maintains itself as an organism," in
Aldo Leopold's words. Normality, of course, ain't what it used to
be. So this "picture of healthy land" isn't a snapshot of a condition,
but rather the guaranteed continuation of a *process*—the process of
evolution unchecked. The baseline may shift, but at least wilderness
can still be one end of that baseline. The wild can serve as a control
against which to measure our myriad experiments in domestication.

In the wild we can have a place where evolution continues
unimpeded (if not uninfluenced), where life adapts on its own, even
if those adaptations are just responses to human actions. When you
think of it that way, Thoreau's declaration that "in wildness is the
preservation of the world" takes on an even deeper and more urgent
meaning. Wildness doesn't merely preserve the world—wildness
perpetuates it.

Reserving some places for evolution can also fulfill the aesthetic
and spiritual dimensions of wildlands. We're a hairless, culture-bear-
ing animal whose biological niche is meaning-making. So symbols
matter, ideas matter—and we need areas that can serve as symbols
of the Away.

The power of a big, bold wilderness like the Arctic refuge can
be seen in the hope people invest in the place. During the political
fights of the early 2000s over oil drilling in the refuge, more than
612,000 people sent messages to Congress in defense of the refuge.
The vast majority of those people knew they would never see the

coastal plain themselves. At some level, they simply wanted to know that it existed. They wanted the assurance that, in a few places at least, what Olaus Murie called "nature's freedom" still reigns.

Make no mistake: exercising restraint will be incredibly difficult. We are likely to witness casualties—a great many of them, no doubt. The pika may perish. The Arctic fox might slip into the great void of extinction. Places we've known and loved—the white-bark pine slopes of the Northern Rockies, to take one example—may become unrecognizable to us, landscapes that no longer fit nicely with our definitions of beauty. To watch such dislocations will require an emotional fortitude to which we are unaccustomed, an almost Buddhist sort of non-attachment.

In the grand scheme of things, it will be relatively easy to keep a place as remote as the Artic refuge free from manipulations. But in choosing wildness there, we can strengthen the muscles of forbearance and restraint for the harder choices to come.

There are much more difficult decisions about global domestication lurking on the horizon. Foremost among those is the question of atmospheric geoengineering. Some scientists are knocking around the idea of manipulating the entire atmosphere or the functioning of the oceans in order to counteract the effects of global warming. One scenario (the most plausible) imagines deploying a fleet of airplanes to spray sulfur particles into the stratosphere to deflect more sunlight back into space—a planet-wide version of pulling down the shades. Another geoengineering scheme would involve dumping huge amounts of iron particles into the oceans to spur plankton blooms; the plankton then gobble up CO_2, die, and sink to the bottom of the ocean, sequestering carbon in the process.

If that sounds like science fiction, know this: some of world's leading climatologists say we need to keep the option on the table, and in February 2015 the National Research Council released a detailed, two-volume review on the feasibility of geoengineering. The report had been commissioned by—wait for it—the Central Intelligence Agency.

Hacking the sky to reverse the effects of climate change? That way madness lies. Geoengineering is a bet that we can save civilization by once and for all divorcing our species from wild nature.

Seizing such ownership of Earth would be a new step in human evolution. It would turn us into a bubble species, living inside a protective dome of our own making. If that comes to pass, we will cease to view the wild world as a comfort, or as our original home. It will have become, instead, a threat.

We are well beyond sentimentality for an unaltered Eden. Claims about "naturalness" won't be able to rebut the drive for geoengineering—certainly not in the post-natural Age of Man. Nor will the arguments about ecosystem services, which have little to say about such a grandiose manipulation. If we want to prevent the ultimate domestication of atmospheric geoengineering, we'll need an iron-clad commitment to the physical and spiritual values of wildness. Such a commitment begins by promising to keep wild the wildernesses we now have.

Long ago, Howard Zanhiser said that defenders of the wild needed to be "guardians, not gardeners." On the shores of the Arctic Ocean, standing next to a washed-up plastic cooler, this avid gardener came to agree.

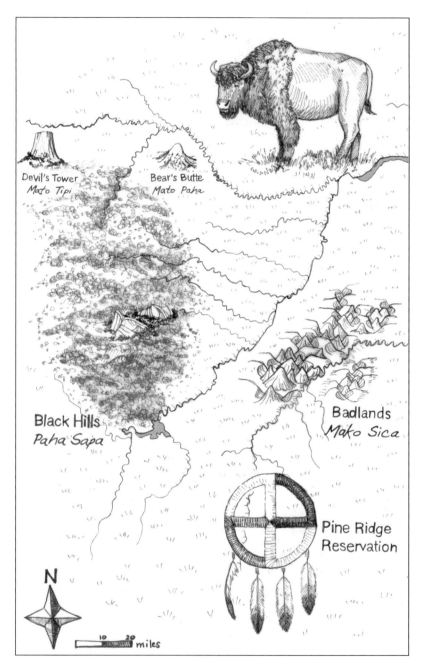

Lakota Country

The Heart of Everything That Is

THE WARNING WAS CLEAR AND HARD TO MISS: a rectangular yellow sign, posted on the barbed wire fence running alongside the gravel roadway of Tribal Route 2. A triangle at the top of the sign depicted a cannon ball exploding into pieces. Below it read

CAUTION

FORMER

BOMBING RANGE

The Area Beyond This Sign Was Used For

Military

Purposes in Past Years

For Your Safety Do Not Disturb Unknown
Objects

They Could Accidentally Explode

Considering myself duly warned, I spread the barbed wire apart, slipped between the rusted lines, and headed into the lonesome reaches of the southern Badlands.

After the endless flatness of the Great Plains, the Badlands comes as a surprise. Sharp, crenellated cliffs of beige, rose, and orange suddenly spring out of the green prairie grasses. Pinnacles and spires twist and fold into wafer-thin ridges, like a meringue made from dust and clay. Hoodoos carve edges into an otherwise seamless sky. It's as if the land had undergone a kind geologic acid bath, erosion peeling away the earth's outer layers to reveal ancient sediments.

The Badlands is a waterless place. Cottonwoods cluster near ravines that are dry most of the year, while here and there a bushy cedar or two hunker in the swales. The main river is called the White, and it runs the color of skim milk. I wouldn't drink it if you paid me. The Sioux call the area *Mako Sica*, which either means "land bad" or "eroded land," depending upon who's doing the translating. The first Europeans to the area, French trappers, took up the name. *Les mauvaises terres a traverser,* they called it—"bad lands to traverse."

The (seeming) barrenness of this (apparently) empty wasteland was one of the reasons why, in 1942, the US military decided the area would make a perfect bombing range. The Army—evidently unconcerned about the obligations of treaty promises—summarily snatched some 340,000 acres of land that belonged to the Oglala Sioux Tribe. The 107 families who lived in the area were given two weeks to evacuate their homes and farms. A few held out, and were given a second two-week warning. Then the bombers started their training flights. They used old cars and yellow-painted oil drums as targets.

A few old timers on the Pine Ridge Reservation still remember how the windows in their homes were blown out by the blast waves.

They also recall, with a mix of pride and resentment, how nearly a thousand Oglala men signed up for the war. Even as the US military was using their homeland as a practice run for Dresden, Sioux were serving as US Navy code talkers in the Pacific or as US Army scouts in Europe, slithering up close to the Nazi lines. (The name *Sioux*, by the way, is an abbreviation of a mashup of French and Chippewa, *Nadouessioux*, which means "little snakes." Except when referring to official tribal agencies, I'll use *Lakota*, which is what the people call themselves.)

The Lakota, of course, are no strangers to betrayal and insults. But even in a history pockmarked with offenses—after the massacres and broken treaties and the carving of four US presidents' white faces into the sacred Black Hills—the seizure of the Badlands for use as a bombing range was an especially sharp fuck-you. The issue still hasn't been resolved. In 1968, the Badlands, including the area inside the Lakota reservation, became a national monument and, in 1976, by act of Congress, a national park. The Southern Unit of Badlands National Park is federal property sitting on Lakota land.

I hiked through the mixed-grass prairie toward a long sandstone formation—just an afternoon reconnaissance to see how far I could wend my way into the arid labyrinth. To the west lay the big mesa of Cuny Table. Somewhere beyond that was Stronghold Table, where the last Ghost Dance occurred and where, after the horror at Wounded Knee in 1890, a force of twenty-seven Lakota warriors successfully held off the US Seventh Cavalry. Whites are prohibited from visiting Stronghold Table; only Lakota go there.

Following the contours of the prairie, I climbed up a draw that appeared to slip through the earth's fortifications. No luck. The way was blocked by a wall of gray chalk. I had no clue how to penetrate the maze. Figuring out such a landscape, I imagined, would take a lifetime of study.

I turned around and headed to my vehicle, which I could spot with my binoculars a couple of miles back. The whine of crickets rose about me like a wall of sound as I threaded my way among the sneaky prickly pears. I was walking across a long shelf of short grasses when I stumbled across the evidence of warfare I had been searching for.

No, it wasn't any long-lost, half-buried bomb. The object was old, to be sure, but its age would be measured in seasons, not years. There were many of them, I now noticed—flat, brown-gray disks scattered randomly, as in a minefield. It's a sight familiar to anyone who has tramped through any part of the American West. From the Northern Rockies to the desert Southwest and from coastal California to the high plains of Colorado and Wyoming, it is probably the most ubiquitous mark you'll find on the landscape.

I kicked the thing with my foot and flopped it over. A dried-up pile of cow shit.

I had gone to Lakota country to explore the most emotionally fraught issue surrounding the idea of wilderness. Where exactly do people fit in our mental picture of wild nature? How can we live with the land in some measure of harmony? How have we done it better (and done it worse) in the past? These questions cannot be understood—they cannot even be approached—without the stories of the people who lived here first, before Columbus stumbled on "so-called America," as one Lakota woman I met on Pine Ridge said.

The dispossession of the Indigenous peoples is the original sin of the United States. The disgrace is so often forgotten perhaps because it's hard to comprehend the enormity of the atrocity. In the space of just a few generations disease wiped millions of people off the map, obliterating entire cultures. The survivors fought as best they could, but inevitably found themselves surrounded by the new settlers. The nations that survived were relegated to the least desirable places (like, for example, the Badlands). Countless others disappeared forever. Today in the United States, Native Americans make up about 2 percent of the population, and their cultures—despite leftover place names and the motifs that some of us decorate our homes with—exist at the far margins of the public mind.

The wrong was repeated during the creation of the United States' much-celebrated parks. "America's Best Idea" is founded on taking land from people, then rubbing out their history. I don't mean that in some polemical, hand-waving way, pressing the uncontestable point that all of this country is somehow conquered territory. I mean the National Park Service (or its preceding agencies) sometimes actually seized the land.

The list is long. Land grabs occurred in the Northern Rockies, where the park service excluded the Shoshone from Yellowstone; and in the Northwest, where Coast Salish names were replaced with English ones, so that a peak long known as "Tahoma" was rechristened Mount Rainier; and around the Great Lakes, where the stories of the Ojibwa who had lived there for centuries weren't included in the "official history" of Apostle Islands National Park. The removal of the Blackfeet from what is now Glacier National Park is one of the best-known examples of Native American dispossession that occurred in the course of making a nature preserve. For decades, park rangers aggressively enforced a ban against Blackfeet hunting and fishing in their traditional territory, despite the complaint by the Blackfeet that the policy violated their treaty rights. During a 1915 visit to Washington, DC, a Blackfeet delegation asked Stephen Mather, the founding director of the park service, to at least keep Blackfoot place-names in the park. A negotiator named Tail-Feathers-Coming-Over-The-Hill said the whites used "foolish names of no meaning whatsoever." Mather accepted their demand. Then he went ahead and used white names on the maps anyway.

Such facts eat at the wilderness ideal like acid. Among all of the critiques of wilderness, the sharpest is the one that faults the Romantics for imagining that the wild could be a place out of time, thereby erasing the people who lived there. "The myth of the wilderness as 'virgin,' uninhabited land had always been especially cruel to the Indians who had once called that land home," William Cronon wrote in his landmark essay, "The Trouble with Wilderness." In the two decades since that article was written, advances in archaeology and paleobiology have given new weight to the criticism. It turns out that humans had spread over just about every part of the hemisphere before Europeans showed up, and that wherever they went they shaped the landscape in fundamental ways. The wilderness that Europeans believed they discovered was, in fact, a human construction.

The new revelations about the Americas before Columbus and the consequences of the contact between the "Old World" and the "New World" have fueled the intellectual debunking of wilderness. "This Edenic world was largely an inadvertent European creation," Charles Mann writes in *1491*, his book about the scale and scope of the pre-Columbian societies. "At the time of Columbus the Western Hemisphere had been thoroughly painted with the human brush." In his book *The Once and Future World*, environmental journalist J. B. McKinnon writes, "To accept that native cultures had the numbers, the knowledge, and power to transform entire continents lays waste to the widely treasured ideal of wilderness."

But *must* it, I wondered? Perhaps there is a middle way, some path that can reconcile the hope that we invest in wild places with the horrors of history. Can we craft a wilderness ideal that is deepened—rather than demolished—by historical awareness?

That's my hope. By instinct and by education I'm more a historian than a naturalist. Whenever I come to a new place—especially if my intent is to hike for days in a big loop and sleep under the stars, as I like to do—I make it a point to learn something about the first peoples who lived there. I've never felt that this diminishes the spirit of the wild. Quite the opposite, actually. Any bit of knowledge deepens my feeling of intimacy with the land, however appropriated that intimacy might be.

In learning how people once thrived in a place, I understand better how the land works. To discover that the original inhabitants caught salmon and made their clothes from cedar bark, to know that they were duck trappers and rice gatherers, to learn how they drove herds of bison off the same cliff for thousands of years, or lit huge fires to construct vast savannahs, or pruned riverside grasses to harvest straight fibers—such knowledge of native economy makes a landscape legible. Story enlarges scenery. It's a way of putting "home" back into the science of ecology.

"One of the penalties of an ecological education is that one lives alone in a world of wounds," Leopold wrote in *A Sand County Almanac*. The line is even truer when it comes to history. One of the burdens of a historical education is that one lives crowded among a litany of miseries. When we restore memory to the landscape, then, we make the wilderness less innocent. The romantic bubble is popped.

All for the best, I say. To put the wild into historical context is to evolve from *scenery*, to *landscape*, to arrive finally at *place*.

Romantics like Muir, so eager for enchantment and escape, made the view into two-dimensional scenery. As we now know, their ideas of the pristine proved unhelpful. Then the ecologists, using the insights of science, turned those flat scenes into full-bodied landscapes. The anthropological explorer can go even deeper by adding the dimension of human experience. Land becomes a palimpsest, stories layered on top of stories, open to multiple readings. To be aware of history transforms a landscape into a place. Which is to say, a site full of memories.

The first memories of America, I am convinced, are still speaking. They have important things to say if we are prepared to listen.

Curtis Temple was the one who told me I would find cow shit in the Southern Unit of the Badlands. He hadn't said it to me in those words, exactly, but he admitted as much when he talked about his grazing leases there. Temple is one of the biggest Indian ranchers on the Pine Ridge Reservation, the 2.2-million-acre spread of South Dakota prairie that the Oglala Lakota were condemned to after their war with the United States. The morning before my Badlands hike, Temple had told me to meet him at the school at Sharp's Corner, near Chimney Butte. I had been waiting there just a minute when he came tearing into the parking lot near the playground. His white pickup was more mud than paint job, and the truck spit gravel as he pulled to a stop. Temple gave me a careful look-over, then invited me to follow him up to his place.

Temple's Pitchfork Ranch sits in a broad swale on the north side of the White River, at the reservation's northern edge. The ranch has been in his family for close to a hundred years, and the care of generations was apparent. The farmyard included a pair of feed silos, a corral and cattle pen, a long, metal equipment shed, and the ranch house, whitewashed and red-roofed and shaded by cottonwoods. We settled into the kitchen, mugs of coffee in front of us, and Temple told me about the land. His great-grandfather had started ranching there. The family fled when the military seized the bombing range, then returned after the war as some (but not all) of the gunnery area was ceded back to the tribe. Many families, he said, never returned.

Now Temple was worried about another land seizure. "My grandparents and my mom and dad and brothers and sisters and hired men, you know, they've all worked hard on this place to get where we are," he said. "I'll be darned if I'm just going to walk away and say, 'Go ahead, government, and run us off.'"

The whole time we talked, Temple kept on his blue windbreaker and black cowboy hat, the band stained with layers of sweat and dust that reminded me of the undulating forms of the Badlands. "It's a land grab, that's what I think," he said. "My main concern is for the reservation not to lose its land. Tribal land, that's our land. It's each and every member's land. Our treaty rights are based on land. If our tribe don't have no land, we aren't going to have no rights whatsoever. Once you lose your land, you'll never get it back."

Temple was upset because, nine months earlier, in December 2013, the tribal government and the National Park Service announced they were canceling grazing leases in the Badlands' Southern Unit, which is co-managed by the Feds and the Oglala Sioux Parks and Recreation Authority. The impending removal of cattle from the park marked one of the final moves in a decades-long process to reintroduce bison there and return full control of the Southern Unit of the Badlands to the Lakota. The handover would create the United States' first Tribal National Park, a preserve both owned and managed by an Indian nation.

When it was first proposed, a lot of Lakota thought that a Tribal National Park would be a great thing. The idea gained momentum in 1999, when President Bill Clinton made a state visit to Pine Ridge (the first president to visit a reservation since FDR) and announced with great fanfare the US government's intention to return the Badlands to the Lakota. At that point, almost everyone on the reservation was in favor of the park plan. People were especially excited about the vision of reintroducing a herd of up to a thousand bison to the Tribal National Park, a proposal that was welcomed as a kind of restorative justice. "This was a very positive plan," a Lakota named Chuck Jacobs would tell me later. "People were saying, 'This is our legacy.'"

But when Lakota ranchers realized that bison reintroduction would mean the loss of their grazing leases on park land, some of them began whipping up public opposition to the whole Tribal National Park concept. Jacobs, for one, said he felt "blindsided" by the campaign against the bison reintroduction. Jacobs had been the tribal treasurer for a couple of years, and from 1997 to 2009 he was the president of the board of the Oglala Sioux Parks and Recreation Authority. He had done more than just about anyone to promote the tribal park. "We've had more public meetings on this than on any other issue," Jacobs told me, by way of making the case that the process had been thoroughly transparent and inclusive. "But then they [the ranchers] got people to come to these meetings, and they pack the gymnasium with screaming people, and it scares these [tribal] council members."

Curtis Temple had been in the vanguard of the opposition. Temple is in his late fifties, with a bushy, graying mustache under a big bulb of a nose. I found him to be an easygoing guy; during our chat at his place, a door slammed with a bang outside and he cried out in mock alarm, "Indians!," cracking himself up. But he was pissed off when it came to the question of the Badlands. "It's just retarded, I think," he said. "It's stupid. The government is trying to take it. They have dangerous wording in the agreement—*adjacent to* and *eminent domain*. They say they won't use it, but they won't take that language out. They're going to wipe me out. That's why I'm fighting this really, really hard."

To push his case, Temple was running for tribal president. The election was just two months away, and driving the roads on Red Shirt Table I saw hand-painted signs reading, "Vote Curtis Temple for OST President." (In the end, he wasn't elected.) Other signs up on Red Shirt Table protested the impending cattle removal—"Ask the Land-Owners!" and "Council Keep Out," the lettering in red on a field of yellow. I spotted those signs while driving past the ranch of the Two Bulls, a prominent Lakota family that had also come out against the park plan.

According to Temple, at least a dozen Indian stock growers would lose grazing rights. Many more people, he said, were opposed to the idea. At this point in the conversation, Temple's ex-wife, Tammy, who had been puttering around the kitchen, chimed in: "It's kind of conflicting," she said, "because they're trying to bring back the buffalo, but then yet they're trying to take the land from people, like the tribe. Trying to take it from a private owner? That's going to be more hard. It's going to be more difficult to take land from the people who are actually standing up and fighting for the land, like we are. The tribal members are really biased on this."

I asked Tammy what she meant by "biased," and she said, "We have half for and half against. We are basically fighting each other down here."

Curtis disputed that estimate, and claimed that well more than half the tribal members were against the Tribal National Park as currently conceived. A big part of the problem, Curtis said, is that buffalo and cattle simply don't mix. Though he was proud of the one bison herd that for decades the tribe has been managing elsewhere on the reservation (the meat used for community gatherings and traditional Sun Dance ceremonies), he pointed out that the bison were a money loser for the tribe. Cattle ranching actually made money.

"The only thing that's been successful on this reservation is ranching," Temple said. "This is just cow country. They tried putting buffalo out there in that [Southern] Unit, and that's pretty much all tribally owned. And they tried running buffalo back there in the early eighties, and they wouldn't stay in there. It's bad water and no grass and the buffalo wouldn't stay in there, just busted through the fences. Hell, we would have buffalo all over this country if the settlers hadn't killed them off years ago. They killed them and just left the meat. All they wanted was the hides. And now they're trying to reintroduce them back. Buffalo and cattle—you have to get a different setup. For buffalo you need stronger facilities, because if you've ever messed with buffalo, they are dangerous."

I asked him if he had ever tried to raise buffalo for market. "Well, no. But I've chased buffalo." I thought of the huge animals I had seen a couple of days earlier on the Badlands North Unit—big beasts, with heads like an anvil and rumps like a linebacker. Temple paused for effect, then said, "And, I've been chased *by* buffalo." Everyone in the kitchen laughed.

It seemed I had stumbled on a new version of one of the iconic struggles of the American West, reenacted for the twenty-first century. "This is just the standard livestock-producers-versus-buffalo," Jacobs told me later. Only there was one major difference from the classic version. Now, some Indians were on the side of the cows.

To try to understand the "Indigenous perspective about wilderness" by visiting a single tribe is, I know, presumptuous. To try to do so through the experience of the Oglala Lakota? That's borderline preposterous.

Is there a more storied tribe in America? The Cherokee, the Iroquois, and the Apache also stake strong claims on the national imagination. The Lakota, however, are the unmatched icon of the American Indian. If you think of riders on fast ponies, garlanded with eagle-feather war bonnets and firing arrows at a circle of wagons, then you're remembering the Lakota and their cousin-allies, the Cheyenne, harassing miners and Mormons on the westward trails. The Lakota heroes—Red Cloud, Sitting Bull, Crazy Horse—are now heroes to people around the world. The famous places of their history are landmarks of Americana. Little Big Horn ("the greasy grass," as the Lakota call it), where Custer's hubris caught up to him. Wounded Knee, where, in December 1890, Chief Big Foot peacefully led his people after Ghost Dancing in the Badlands, only to be gunned down by US troops, the bodies left to harden on the frozen ground.

Today at Wounded Knee there's a small cemetery on top of a knoll. The tombstones are numbered and many-named:

32 Wolf Skin Necklace

33 Lodge Skin Knowing

34 Charge at They

35 Weasel Bear

36 Bird Shake

37 Big Skirt

. . . and so on, for some of the at least 200 men, women, and children who were killed there. Brightly colored prayer flags tied to a chain link fence whip around in the wind. There used to be a museum, but it's been abandoned. The day I visited, a couple of guys in a Camry were selling trinkets on the roadside. I gave them a bill with a portrait of Andrew Jackson—archenemy to the Seminole and the Cherokee—and they gave me a necklace with a white buffalo calf, a sacred symbol.

Contested terrain, I wrote in regard to Point Reyes National Seashore and the Arctic National Wildlife Refuge. Those are mere squabbles. Lakota country is a real, blood-stained battlefield, one that is burdened and bedeviled by myth.

But I didn't travel to Pine Ridge just because of the Lakota's charismatic history. I ended up there naturally, a journalist on the hunt for a good story. By 2013, the planned return of the South Unit of the Badlands to the Lakota (about 133,000 acres total) and the proposed bison reintroduction had dribbled out into the media. "In the Badlands, a tribe helps buffalo make a comeback," the *Washington Post* reported. "Good News from the Badlands," an airline magazine trumpeted. National Park Service officials were beginning to talk up the deal. "I think, rightfully, the South Unit should go back to the tribe," National Park Service Director Jon Jarvis told me in a June 2014 interview. "Because we've managed [the South Unit] so long as part of Badlands National Park, the opportunity to establish really the first Tribal National Park in the United States is very, very powerful."

"This is really unique. Nothing else is like it," Eric Brunnemann, superintendent of Badlands National Park said to me. Brunnemann is a trim, angular man, with the crisp bearing of an Air Force lieutenant and the sincerity of a Boy Scout. While he was careful not to get too excited about sealing the deal, he couldn't contain his hopefulness. "It is an incredible story. This is one of those cases where we don't know where we are in the story. Are we at the end? Are we at the beginning? We don't even know how far into the book we have gotten, but it really is an exciting read. I think it has great potential to be something that's discussed for years to come."

I, too, was eager to see where the story would lead. I was hopeful that I would find some tale of a wrong made partly right: land returned to the locals, bison returned to the land. Turns out I was overly optimistic. As I had begun to understand from talking with Curtis Temple and Chuck Jacobs, the Tribal National Park had run into a wall of controversy.

"The people don't fucking want it," Temple had told me. Indeed, the more time I spent on the reservation, the more it became clear that he was right: most Lakota were skeptical of the Tribal National Park.

In Chuck Jacobs' telling, though, the public uproar was a phony debate that had been ginned up by the cattle ranchers. "They have done a good job of disinformation," he said. "There are so many outrageous conspiracy theories out there. People talking about oil and mines—the Badlands are the largest zeolite deposit in the world. But this is *added* protection." We were sitting in my rented Subaru Outback in the parking lot behind the Bureau of Indian Affairs' brick building in the center of the town of Pine Ridge, the largest settlement on the reservation. The September sun was baking the asphalt, and I was sweating, and so was Jacobs, who was wearing black cowboy boots and black jeans and a white straw cowboy hat. Taco John's or Subway (two of the few restaurants in Pine Ridge) probably would have been more comfortable, but when we were

arranging a place to meet I got the sense from Jacobs that he didn't
want to be overheard when talking about the Badlands.

Jacobs said the public had it all wrong. The US government
wasn't taking the land; the US government was *returning* full control
of the territory to the tribe. "No private lands or homes will be
taken. One of the ranchers out there"—Jacobs later clarified that he
meant Temple—"he's one of the biggest ranchers on the reservation. He can afford to be downsized. This project isn't going to hurt
them. These guys are fighting over fifty-head units. That family is
just used to being cattle barons. This is a classic case of the benefits
of the whole outweighing the benefits to the few."

To an outsider, it was baffling. I couldn't help but get the feeling
that what should have been a small misunderstanding about bureau-
cratic legalese had metastasized into full-blown mistrust. But suspi-
cion grows easily in the soil of injustice. Eventually I would come to
learn that the fight over the Badlands was incomprehensible without knowing the Lakota's history—above all, the way in which the
Black Hills had been taken from them.

For Jacobs, the resistance to the proposed park plan represented
nothing less than an identity crisis. It was "so unbelievable" that
"because of all this internal strife, these guys don't even realize what
they are doing. They are saying *No* to the buffalo." Jacobs laughed
and shook his head, stumped with incredulousness. "All the other
Lakota bands in South Dakota can't believe we are saying No, that
we are considering not doing this project with a thousand head of
buffalo. I mean, *who are we?*"

⁂

George Catlin—the 1830s-era conservationist and painter whose
canvasses of the Great Plains and their inhabitants helped establish
the popular imagery of the Wild West—called the Lakota "the dancing Indians" because of their tireless, days-long ceremonies and cel-
ebrations. From what I saw during three days at the Porcupine powwow, the name holds true.

In the weeks before my trip to the reservation, I made dozens of
calls to try to schedule meetings. Invariably I was told, "Just call me

when you get here." Finally, one guy gave me a tip: "Oh, you'll be here on Labor Day weekend. That's the Porcupine powwow. It's the last one of the summer. Everyone will be there."

The community of Porcupine (*Pahin-Sinte* in Lakota—literally, "spiky tail") spreads across a north-south valley flanked with pine-frocked buttes. In a streamside meadow off the main road sits the town's ceremonial arbor. It's a wide, round, open-air structure with a painted plywood roof and lodgepoles standing against every other interior post. The entrance is on the east side, facing the morning sun, just as the Lakota traditionally oriented their tipis. Families set up lawn chairs underneath the arbor to watch the action. Behind the arbor a loose collection of trailers and pop-up stands were selling soft drinks, fry bread, and burgers.

From 11:00 a.m. until dark the dancing never stopped. There was the "fancy dance," a fast-paced, frenetic choreography of twirling and spinning, dancers hopping from foot to foot while going in circles. There was the "shawl dance," which is the women's version of the fancy dance. It's just as speedy but more stately, the women spreading their arms wide with each turn, winged tunics making them look like birds. My favorite was the "grass dance," which seemed to best combine power and grace. Male dancers, wearing ribboned pants and tunics, bounced around the circle, making quick little steps and crouching low, as if sneaking up on a pronghorn.

Except for short breaks between dances, the drumming was constant. All day long a constant *thump-thump-thump-thump* that occasionally exploded into adamant bass booms. The men sang as they pounded the drums—a piercing, high-pitched quaver almost like an ululation, the words (as best as I could follow) looping again and again with insistent rhythm. A jovial emcee filled in the moments between dances, supplying a jokey commentary in a mix of Lakota and English, like a Sioux Bob Hope.

A dude named Rider helped interpret the scene for me. Rider, a thirty-one-year-old, half-Lakota, half-Cheyenne guy from Wounded Knee, and his wife, Olowan (almost rhymes with "aloha"), own a burger-and-fry-bread stand called Fire Lighting Concession. We hung out together for much of the three-day gathering, sitting in Rider's canvas concert chairs and smoking rollies from my pack

146 SATELLITES IN THE HIGH COUNTRY

of Drum tobacco. After the Indian Wars, Rider told me, the white
authorities cracked down on the Lakota's traditional dances as part
of the larger push for Christianization. For decades, the dances
could only occur in secret. Then, in the 1970s, at the height of the
American Indian Movement's wave of activism, the powwows and
Sun Dances were revived and brought back into the open. "When
I see the young kids like that," Rider said during the tiny-tots-
division dancing, "it makes my heart feel good, knowing that it carries on. These ceremonies were brought to us by our great-great-
grandfathers and -grandmothers. It's part of who we are. It's in our
blood, in our DNA. It's how the Creator, *Tunkasila*, made us."

While proud of their culture's resilience, neither Rider nor
Olowan made any effort to gloss over the dire conditions on Pine
Ridge, which Olowan referred to as "the injustices all around." The
reservation is an infamously poor place, typically ranked among the
poorest locales in the United States by the Census Bureau. More
than half of people live beneath the poverty line, and the per cap-
ita income is $8,700 a year. Infant mortality is high—300 percent
above the national average. Life expectancy is low—on average men
live to 48, and women to 52, the lowest averages in the Western
Hemisphere outside of Haiti. Obesity, diabetes, and heart disease are
endemic. So is alcohol abuse and drunk driving, which seems to be
a masochistic blood sport on Pine Ridge. Olowan called alcoholism
among the Lakota a "stillness of the mind, a liquid genocide."

Rather than succumb to such statistics, the couple had committed to resisting them. "We still have that fighting warrior spirit,"
Olowan said. She and Rider described themselves as "AIM babies,"
and said that spending the summer on the "powwow trail," going
to tribal gatherings across the West, allowed them to keep in close
contact with other members of something called the Native Youth
Movement. As part of their activism, the couple had been involved
in civil disobedience protests against the beer distributors in White
Clay, Nebraska, a town on the edge of the reservation that basically serves as a giant liquor store. More recently, they had gotten
involved in the fight against the Keystone XL pipeline (an issue still
very much in play in the fall of 2014, when I was there) and what

Olowan described as the "environmental warfare" occurring against native communities.

"We're standing up for the land and the water and standing up for Native rights," Rider told me. He's a round man, keeps his hair in a long, thin ponytail, and both he and Olowan were wearing black hoodies with the Guy Fawkes mask. Only they had customized their sweatshirts: a lightning bolt, painted in the red of the tribal flag, zigzagged from the top of the mask to its chin. Rider said, "Water has always been sacred to us. It's the first medicine. Without water, everything will die. All of this oil—all of this fracking—it's hollowing out the earth. I can't wait until *Unci Maka*, Grandmother Earth, opens and swallows them up for what they are doing."

Toward the end of the powwow's second day, the drumming stopped and the crowd got silent as a group of riders came down the bluffs to the west. There were probably twenty-five of them, mostly teenagers, with a couple of adults at the front and the rear. Some had saddles and bridles, but many more rode with just a horse blanket and halter. A few were bareback. I saw a pair of teenage girls ride by on a dappled gray, the girls' long, black hair matched to their mare's dark mane.

A man led the riders into the entrance of the arbor. The announcer introduced him as Percy White Plume and said they had come over the hill from Manderson. Sitting on his horse, White Plume gave a short speech, some of which I managed to scribble down. "*Tiyospaye*," he began, addressing himself to his extended family, his clan. "This is beautiful land. People say the reservation is a bad place. But we have to remember and recognize that we have a beautiful reservation. I bet you've been on this road and looked at those hills and have wondered what is over there. We rode cross-country from Oglala to here. No tracks. No roads. Just out in the open. The kids are taken away from their computers and cell phones. No Internet. So they see just how beautiful our land is."

I knew what White Plume meant. Pine Ridge was one of the prettiest landscapes I had ever seen—"God's country," a woman at Rider's stand volunteered to me. It had been a wet summer in the Dakotas, and the prairie grasses were a swirl of green and gold, soft as a watercolor wash. Sandstone buttes capped the hills while bottomlands of box elder announced where the streams lay. Many Lakota still keep horses, and often I spotted small herds of Appaloosa grazing in fields of tiny, wild sunflowers. Almost every afternoon a thunderstorm came over the Black Hills, split the air with rain and lightning, then passed eastward, leaving the sky looking tousled. At night the stars were as sharp as in any wilderness. "On Pine Ridge, dark is darker, somehow," Ian Frazier writes in his tender profile of the Lakota, *On the Rez*. It did feel like a place of unexpected charms.

I share all of this as a way of pointing out that Lakota ideas about the land—about how we two-leggeds are supposed to share space with the four-leggeds—are not merely memories, historic relics, or anthropological curiosities. The Lakota Way remains a living worldview—marginalized, to be sure, but still standing. The Lakota may have surrendered, but they were never fully subjugated. The US government's stated goal had been to destroy their culture. "We must act with vindictive earnestness against the Sioux, even to their extermination—men, women, and children," General William Tecumseh Sherman wrote in 1866. The Lakota had suffered shootings and bombings, seen their children shipped off to schools back East where their culture was whipped out of them. And yet they endured. By some important measure, the Lakota won the war. How many European-Americans still practice the traditional dances of our fathers' forefathers? Well, the Lakota still do.

William Faulkner famously wrote that the past isn't over—it isn't even past. In much the same way, it seemed that the Indian Wars are not really finished. The Lakota are still very much fighting, though in a new, twenty-first-century fashion that involves rallies on the National Mall to protest the Keystone XL pipeline or sit-ins to block equipment headed to the tar sands mines from crossing their lands. "Grandmother Earth is tired of this, tired of people tapping her veins, and I'm tired of it, too," Olowan told me one night as the powwow was winding down. "How far are we going to let them stomp on what's sacred? They've already been doing it for hundreds of years."

For Olowan, it all came back to the question of the land—who it belongs to, and how *we* belong to *it*. When Rider introduced us, the very first thing that Olowan said to me upon noticing my reporter's notebook was "It's been about land ever since the Invasion—ever since Day One."

Often when I spoke with people on Pine Ridge, it sounded like they were channeling Nicholas Black Elk, the famous Lakota holy man. His autobiography, *Black Elk Speaks* (which he dictated to a white man, John G. Neihardt, in the 1930s), is a classic of religious literature and a touchstone for Lakota spirituality today. The book tells the story of the powerful visions Black Elk had as a boy, which started him on a path to becoming a healer and a spiritual medicine man. Black Elk speaks of "the beauty and strangeness of the earth" and of how "every little thing is sent for something." He is like a Great Plains St. Francis of Assisi, refusing to join the boys of his youth in throwing stones at little birds because "the swallows seemed holy."

This image of the American Indian living in harmony with wild nature is, of course, the classic Indigenous archetype, and it's an ideal that continues to anchor Lakota identity. Wherever I went I heard Earth-centric talk.

"We are people of the land," Tom Poor Bear, the tribe's vice president told me. He was wearing a black leather vest and a bandana that was cinched with a bone-bolo painted in the Lakota's four-colored medicine wheel. "We have only one mother—and that's the earth."

A man named Wilmer Mesteth said, "We don't over-harvest. We take what we need and then we move onto another area so the plants can regenerate. Same with hunting." Though he seemed resistant to the title, everyone I met referred to Mesteth—who teaches native plant biology and Lakota pharmacology at the college—as a medicine man.

Karen Lone Hill, chair of the Lakota Studies department at the tribal college, said, "We see Earth as a living organism. If you take care of the land, it will take care of you."

Yet when it came to the particulars of the Badlands, the idea of stewardship got more complicated.

"I believe in land and animals, too," Curtis Temple told me when we were talking in his kitchen. "But we don't need someone coming in here and saying something's endangered to take our land."

"I'm all for reseeding the buffalo, but not where they're going to take people's property," Mesteth said.

To be sure, every modern person is hamstrung by inconsistencies when it comes to our relationships with nature. And the Lakota, although they may be half a step closer to the old ways, are no different. At the Porcupine powwow, a former tribal council member named Mike Her Many Horses gave me an earful about the Keystone XL pipeline. Then he admitted with a grin, "But I *do* love driving." As do most Lakota. The long, straight roads of the reservation are made for pushing the gas pedal, and I was told it's not unusual to put 40,000 miles on a vehicle in a year.

The contradictions I perceived between the Lakota's self-image as environmental champions and the mess of controversy surrounding the Tribal National Park seemed to echo the emerging scholarship about pre-Columbian landscape management. As new research shows, the stereotype (what else to call it?) of the American Indian living in perfect accord with nature isn't entirely accurate. The Lakota, like most other Indigenous peoples, lived in a wild that they cultivated.

In an 1872 petition to Congress, the legislature of what was then the Dakota Territory complained that the Lakota were making "no legitimate use" of the Black Hills, instead only using it "as a hiding place to which they can flee after making depredations." The rhetoric reflected the standard white man's view of the time. Here was this vast, virgin territory that the Indians were letting go to waste because they were not felling its forests or plowing its soils. Of course, the Indians *were* using the land—quite intensively, actually—but in a manner that was too subtle for European-Americans of the time to understand.

A new body of archaeological research has demonstrated that Indigenous cultures across the Americas shaped their environments in fundamental ways. "We cannot assume that [Native Americans] always walked lightly on the Earth," anthropologist Sheppard Krech III writes in his provocatively titled book, *The Ecological Indian: Myth and History*. Indigenous Americans weren't living some kind of Rousseau fantasy—children of the earth, gently plucking the ripe fruits. They acted as sophisticated ecosystem managers, crafting the world they lived in.

From the moment *Homo sapiens* passed over the land bridge from Asia and began to spread across the Americas, people were changing the land—and not always for the better. An astounding range of large animals were the first casualties of humans' arrival. Fifteen thousand years ago, the Americas were populated by all sorts of animals that no longer exist. Mastodons roamed the plains, along with huge dire wolves, saber-toothed cats (the smilodon), and massive, ten-foot-long armadillos called glyptodonts. There were ostriches on the plains and humongous beavers in the forests, plus giant ground sloths that could reach branches twenty feet in the air. (If you've ever wondered at an avocado's huge pit, it likely coevolved with the large digestive tract of such beasts.) Then this amazing bestiary disappeared. The die-off was swift—many large species went extinct between about 11,500 and 10,900 BC—a time period not long after the appearance of people.

Scientists continue (often angrily) to dispute the degree to which humans contributed to this Pleistocene megafauna extinction. Climactic changes might also have caused the animals' disappearance, which came toward the end of the last ice age. But few researchers question that people must have played some role. Combine a steady human population increase with the fact that most of the large mammals had a slow rate of reproduction (a mastodon gestation was about twenty months), and it's not hard to imagine how people could have wiped out a range of species without noticing until it was too late. In short: small bands of hunters, armed only with spears and fire, transformed a continent.

Over millennia, humans in the Americas wrought even greater changes on the land as they manipulated environments to suit their needs. The most obvious examples are the complex empires in Central America and the Andes. But there were also large, sedentary, agricultural societies north of the Rio Grande. The Mississippi Valley "mound builders" of Cahokia and the expert irrigators of the Sonora Desert, the Hohokam, being the best-known cases. And just like larger civilizations, such as the Maya, that outgrew the carrying capacity of their environments, the North American Indian cultures sometimes ended in ecological collapse. At one point, as many as 50,000 Hohokam might have farmed the broad valley that is now Phoenix; by the year 1400 no one was left. In the language of the Pima Indians, Hohokam means "all used up."

Even the nomadic Lakota, although they disdained farming, were engaged in large-scale ecosystem management. They accomplished their terraforming through the most ancient of human technologies: fire. On the upper Missouri River, a white visitor recorded in 1805, the Indians burned the grasses around their village every spring "for the benefit of their horses." One settler wrote of conflagrations so huge they "ran with great velocity & burnt with very great fury, which enlightened the night like day."

The Indian practice of using fire to shape the landscape was widespread. Sometimes, as in the hardwood forests of the East, tribes used fire to make way for their fields of corn, beans, and squash. Most often, it was used as a hunting technique: either to flush game from the brush, or—more impressively—to create savannah-like groves where it would be easier to stalk prey. The first settlers to the Ohio River Valley commented on woodlands in which the trees were spaced widely enough to drive a buggy through. An early-eighteenth-century European visiting the Southeast wrote, "Hundreds of Indians [spread] themselves in Length thro' a great Extent of Country," then "set the Woods on Fire" to hunt deer.

In California, Indians also routinely set fires, both as a game-management strategy and to encourage the growth of seeds that they harvested. The Yosemite Valley that John Muir fell in love with was a human construct. The Ahwahnechees had been firing the landscape for as long as they could remember. In the 1920s, the "last Yosemite Indian," a woman named Totuya (aka Maria Lebrado), who was Chief Tenaya's granddaughter, returned to the valley for the first time in decades. She didn't like what she saw. The place was "too brushy" and "dirty." Without fire—without a human hand on the landscape—the place had become unkempt.

The anecdote reveals the extent to which European-Americans saw the landscape they wanted to see—which is to say, unpeopled. And they were able to see the New World that way because the North America they came upon was, even by its own recent standards, relatively unpopulated. By the time most Europeans arrived, their diseases had already emptied out the land.

The estimates of pre-Columbian human population of North America are fiercely contested among researchers. The numbers range widely, from as low as 2 million to as high as 8 million, with the population of Mexico being as high as 20 million, if not more. (There are distinct bands of disputing academics, the High Counters and the Low Counters.) Even to go with a middle guesstimate of around, say, 4 million people in North America in 1491—which comes from William Denevan, a geographer at the University of Wisconsin—is to acknowledge that humans were more widespread than previously thought.

The land only appeared empty because epidemics had raced ahead of the colonists. A single contact between a Native and a European infected with smallpox or influenza was enough to spark a viral apocalypse—a cursed meeting between, say, a Wampanoag trader and a boat full of sickly European fishermen plying the North Atlantic cod stocks. Some of the very first English settlers to Massachusetts Bay found skeleton-filled villages, disease having already taken its toll. (One early colonist called the spread of disease a "marvelous accident"—holocaust by happenstance, as it were.) In the middle of the continent, the Lakota were suffering from smallpox twenty years before they crossed paths with Lewis and Clark. A Lakota "winter count" of 1784 (the tribe's method of marking history) features a horrifying pictograph: a pox-scarred man, alone in his tipi, killing himself.

Essentially, the Old World diseases created the wild landscape of the Europeans' imagination. In puritanical New England, Cotton Mather's "howling wilderness" might very well have been more howling than it had been fifty years earlier; the diminishment of the Native inhabitants opened the way for wolves to increase in numbers and range. The horizon-spanning herds of bison that blew explorers' minds might have been an ecological anomaly, the result of a collapse in human populations.

The historical revisions about Native societies' environmental impacts are important. Above all, they explode some often-racist ideas about cultural superiority. The historical corrections return agency to the peoples who originally occupied this land. The new findings flip the tired stories about "savages" into a more interesting tale of sophisticated ecosystem managers.

But the correction may have become an *over*correction. To paraphrase geographer Thomas Vale, we may have simply replaced the myth of the pristine wilderness with a new myth of the humanized landscape.

Most often, the new understanding of Indian ecosystem management is used to wave away the importance, even the existence, of the wild. I hear this argument all the time. Nearly two years after the oyster controversy in Point Reyes was all but settled, a writer in the local newspaper was still hammering away on the argument that "wilderness . . . is a romantic fiction" because, supposedly, "North and South Americans have been living in it [the Anthropocene] since well before Europeans made permanent settlements here." Ironically, the argument reinforces the very pristine-profane dichotomy that it is intended to rebut, since it assumes that all human uses of nature are necessarily destructive, and once again conflates the untouched and the untamed.

At its most pernicious, evidence of Native Americans' ecosystem interventions are used as a cynical "gotcha" to justify the kinds of environmental manipulations that would boggle the Indian mind. For example: a prominent commentator like Jonah Goldberg writing a nationally syndicated column that makes a case for oceanic geoengineering because "the pristine natural world has been gone for a long time," and we better "get used to it." Goldberg accurately points out that American Indians "cultivated plants, cleared forests with extensive burning . . . and otherwise altered the landscape." But he then takes the inference too far, and argues that there's no sense getting worked up about dumping tons of iron filings in the ocean to spur plankton blooms. "Ideological" complaints about transforming natural systems are "ridiculous," Goldberg argues. "That ship sailed at least 10,000 years ago."

To compare pre-Columbian use of fire with planetary-scale geoengineering beggars credulity. It elides the fact that Indian land management—while certainly still management—was qualitatively different than what industrial society routinely does today. The question is not whether Indians managed the land—of course they did!—but *how* and *to what degree* they managed it.

Yes, Indians were systematically "tending the wild," to borrow from the title of a book about the Indians of California by M. Kat Anderson. Because of its fresh insight, people like to focus on the word *tending* in the phrase. What if, instead, we focus on the word *wild*? That might lead to a different conclusion—a remembrance that even as Indians shaped wild nature, they also accommodated themselves to it.

One evening after the day's routine thunderstorm I went for a hike at a place in the Badlands called Sheep Mountain Table. Stands of cedar speckled the mesa top, and as I walked through the prairie full of goldenrod and plains sunflower, the end-of-day orange cast the Black Hills into silhouette. The noise of a million crickets and grasshoppers came at me in surround-sound, an incessant chorus broadcast in at least a half-dozen tones: a turntable scratch, a constant trill, a stutter-stop clicking, and others that I'm sure went unheard.

Walking the trail I spotted several different kinds of animal scat—the easily recognizable pellets of deer as well as the lumps from a bobcat and the twisted leavings of coyotes. Common sights, as ordinary as shit. Yet the animal sign helped me think more clearly about how humans have used the world in the past.

To be sure, a human footpath is a shaping of nature. But it's also a *sharing* of nature. Coyote, deer, and bear can, and often do, use humans' backcountry trails, just as the Plains Indians followed well-worn routes trampled by bison. A trail through the wild accommodates multiple users. An interstate highway—not so much. Any animal that tries to travel blacktop quickly ends up as roadkill.

The difference between a footpath and a highway reveals the importance of including scope and scale—and, especially, *intention*— in any conversation about Indian uses of the land. Without question, Indian cultures changed the landscapes they lived in. They experienced nature not just as scenery, but as sustenance. At the same time, most of their changes didn't halt the recurrence of other natural processes. Indians disturbed the land, but did not strive to exercise dominion over it. The world most Indians lived in remained essentially wild and self-willed.

Is there any doubt that the experience of living cheek by jowl with wildness influences people's worldview? This would have been especially true of nomadic tribes like the Lakota that didn't embrace agriculture. (The Lakota looked down on the neighboring Mandan, famed for their gardens and cornfields.) Among other things, the wild life of the nomad shaped an understanding of time that is fundamentally different from the European-Christian view.

"You see, we Indians live in eternity," Ella Deloria, a Yankton Dakota linguist and ethnographer, once said. For the Lakota, time wasn't a straight line, but instead a circle. In traditional Lakota belief, creation is not a singular moment—the epic eruption of Genesis— but an ongoing, everyday process. This sense of time reflects Earth's own logic of cycles within cycles: the pulse of sunrise and sunset, the rhythm of the moon, the repetition of the seasons. And it was through their esteem for the circle that the Lakota (like many other Native cultures) developed a complex understanding of the obligations that we humans owe to the rest of nature.

The Lakota medicine man Nicholas Black Elk spent a lot of time talking about the virtues of the circle. "There can be no power in a square," he says, complaining about how the whites tried to make the Lakota ditch their tipis and live in log cabins. For the Lakota, the preeminent holy symbols are the four-colored medicine wheel and the sacred hoop. Black Elk said, "You have noticed that everything the Indian does is in a circle, and that is because the Power of the World always works in circles, and everything tries to be round."

As with many belief systems, cosmology shapes ethics, including the Lakota's moral standards as they apply to the nonhuman. "We know that we are related and are one with all things of the heavens and the earth, and we know that all the things that move are a people as we," Black Elk said. The circle is a way of expressing the interlocking relationships among the human-people and the buffalo-people and the wolf-people (as the Lakota traditionally addressed other species). The circle creates bonds.

"The greatest principle the circle symbolizes for me is the equality that applies to all forms of life," a modern Sicangu Lakota scholar, Joseph M. Marshall III, writes. "In other words, no form of life is greater or lesser than any other form. We are different from one another certainly, but different is not defined as 'greater than' or 'less than.'"

Another modern Indian writer, the Potawatomi biologist and essayist Robin Wall Kimmerer, makes similar points in her beautiful and sensitive book, *Braiding Sweetgrass*. She points out that most Indigenous languages use the same pronouns for both animals and people. And this, she says, can help us understand "the natural world as a member of the democracy of species, to raise a pledge of *interdependence*." Her emphasis there underscores Kimmerer's larger point: to live in a multi-species democracy, like any democracy, involves mutual obligation and accountability. Native cultures were cultures of gratitude. And, she says, "cultures of gratitude must also be cultures of reciprocity." The Indians took from the land—and they were also committed to the idea that they had to give back somehow. As Kimmerer points out, balance is not a passive act; it takes effort and intention.

The difference between traditional Indian cultures and today's industrial culture isn't a question of use—but the *type* of use. The distinction mostly hinges on the commodification of nature. As soon as an animal or a tree or a piece of land is commodified, it is transformed from a *being* to a *thing*. Its intrinsic worth is turned into instrumental value. Giving and borrowing are replaced by taking.

Eventually, of course, the Indians got sucked into this system of thinking. For all of their eco-rhetoric, there's no question that Native Americans were complicit in the near-extinction of the beaver and the buffalo. Once the white traders showed up, a beaver pelt was no longer a beaver pelt, but something that could be exchanged for gunpowder or sewing needles. A buffalo cow (its hide the softest and most pliable, destined to become a machine belt in the factories of England) was transformed from a walking feast into a trade good.

It has been well documented how the American Indians' sudden injection into the global market economy changed their habits of mind. Still, the old ways die hard. As I was learning on Pine Ridge, for the Lakota some things remain so sacred that they are simply not for sale.

The more I talked to folks about the Badlands and the proposed Tribal National Park, the more the situation bewildered me. I could understand many of the economic concerns. Some Lakota were worried about how splitting the Southern Unit from the National Park Service would cost the tribe its share of the park's gate receipts—about $1 million a year. Others were concerned that the tribe would surrender up to $250,000 in annual grazing fees, or questioned how the tribe would pay for miles of fencing to keep the buffalo contained.

These seemed to me the kinds of matters that could be worked out with patient deliberation. The National Wildlife Federation had promised to pay for the buffalo fencing. In time, the gate receipts might be made up if the tribe did a good job of promoting the United States' first Tribal National Park. As for the legalese that everyone was worried about—the lines about "condemnation" and "adjacent to"—Chuck Jacobs said the wording could be rewritten. "It's a simple fix," he told me. "Take that language out and have an amended ordinance."

The most baffling thing was the sheer intensity of the opposition. It seemed to me that people were tilting at windmills that just didn't exist. Everywhere I went, I heard the same heated talk.

"I believe in the buffalo, don't get me wrong," tribal vice president Tom Poor Bear told me as he left a sensitive meeting with

Badlands superintendent Eric Brunnemann and other NPS officials. "But I believe in our people and I believe in the survival of our people. A lot of this land that is going to be taken away, it has been passed on from generation to generation."

Over coffee at Big Bat's—a one-time French trapper's trading post and now a Texaco gas station that is the reservation's social hotspot—a columnist for the *Native Sun News*, Jeff Whalen, said, "The people do not trust the government. Period. Not after all of the land-stealing that has gone on. It hits a button, probably the most sensitive issue in Lakota consciousness."

Even Harold Salway–Left Heron, a two-term tribal president and now the executive director of the tribe's parks and recreation authority, an agency that would apparently benefit from the handover, was against it. "Our society emanated from the buffalo, and they [white settlers] eradicated the buffalo knowing it was our staple food," he said. "Looking out for the buffalo is like looking out for one of our own close relatives." And yet: "The Park Service promoted the concept of the Tribal National Park, but it happened without full and continual input from our people. To me, it's a land grab. . . . The concept of 1,000 head of buffalo in the Badlands is out there."

(In October 2014, a month after I visited Pine Ridge, the tribal council rescinded the ordinance that canceled the ranchers' grazing rights and voted to hold a tribal referendum on the question of the Tribal National Park. As of May 2015, the referendum had not yet been scheduled.)

A former AIM activist named Milo Yellowhair was the one who finally broke it all down for me: the Badlands situation had become embroiled in something bigger. The removal of a few ranchers' grazing leases had become a symbol of past treacheries.

I happened to catch Yellowhair talking one morning on the reservation radio station, KILI. "Is there room for sacredness in the American Dream?" he asked his listeners as he discussed the ongoing campaign to recover the Black Hills, the forested range to the west of Pine Ridge that the Lakota never formally surrendered. "We need to remember that we can have a sacred connection to the

land. Without that, we are rootless." Intrigued, I called the radio station and eventually tracked down Yellowhair. We managed to meet
on my last day on the reservation, and he laid out his take on the
controversy.

"No matter how you look at the situation, it has its roots in
American policy," he said. "I always quote a guy by the name of
Little Wound, who said, 'Goddamn the surveyor.' Because when you
try to fit a square into a circle, you leave a lot of things out."

We were sitting in a borrowed pickup truck, the cab piled with
neat stacks of magazines and newspapers: *Time*, the *Economist*, the
Wall Street Journal. For years Yellowhair was the radio station's news
director, and in the nineties he served a stint as the tribe's vice president. His jet-black hair hung past his shoulders, framing a round
face and thick, square glasses. "Native Democrat," his T-shirt read.
To Yellowhair, the Badlands fight was of a piece with the long-running Lakota campaign to reclaim the Black Hills. Both controversies revealed fundamental differences in how Native peoples and the
dominant white culture view wild nature.

"This is something that began with Cain and Abel. One was a
farmer and the other was a nomad. And the farmer ended up killing the nomad. And it repeats itself. It's a common thread through
history. The inherent conflict between the farmer and the nomad
must be settled through peace—but the farmer is always encroach-
ing. The greater society, the dominant society, has to accommodate
itself to living in America, because they are still strangers in a strange
land."

I asked Yellowhair what accommodation would look like. "The
idea is to not treat land as a commodity, but as home," he said. "Try
to understand that land. Because every nook and cranny of this
beautiful land that has been the United States has an Indigenous
story attached to it. And there are songs that are attached to that
particular piece of land. There are ceremonies that are attached to
that particular piece of land. There are people who lived there for
millennia, that lived and loved on that land."

The fight over the Badlands, I now saw, was inseparable from
the much older battle over the Black Hills—a conflict that, among
the Lakota, carries a talismanic weight. *Paha Sapa* is the name of

the Black Hills in Lakota. The hills are also sometimes referred to as *Wahmunka Oganunka Inchante*—"The Heart of Everything That Is."

Traditional Lakota myth says that the Lakota people were birthed from the Black Hills. According to their creation story, the first people came from deep within the earth and entered this world from the tiny mouth of a huge cave. (This is now Wind Cave National Park, due west of Pine Ridge, so named because pressure differentials between the caverns and the surface lead to steady gusts from the cave entrance.) According to Lakota tradition, the bison also originated from Wind Cave, emerging "like a string of tiny ants." Once they took in breath, the animals expanded "into their natural sizes."

Ethnographers tell a different story. They say the Lakota migrated onto the Great Plains from the woodlands of the Upper Mississippi and the Great Lakes in the late 1600s due to constant warfare with the Assiniboines, Crees, and Ojibwas. Once on the plains, they developed a unique culture centered on the horse and the buffalo. Eventually they made their way across the flatlands to the Black Hills and the Powder River Basin of Wyoming, where they displaced the Crow while creating an alliance with the Cheyenne, who had long lived in the Upper Plains.

When I left the reservation I headed to the Black Hills to experience the place for myself. I drove by way of Red Shirt Table, over the tiny headwaters of the Cheyenne River, through Buffalo Gap National Grassland, and into the wooded uplands. That is, I came to the Black Hills the way the Lakota would have, from the east. "A forested island in a sea of grass," is how historian Jeffrey Ostler describes the area in *The Lakota and the Black Hills*, "one of the most extraordinary landscapes of the Great Plains." It's easy to imagine that, to a people accustomed to flat woodlands and the plains, the hills would have would have been like a revelation—a dramatic eruption of the earth, right where the sun sets.

To get a feel for the land I spent three days and two nights backpacking through the Black Elk Wilderness, a tiny designated wilderness south of Mount Rushmore. It's impossible to overstate how different the hills felt after the grasslands. To the Plains Indians, the lushness alone must have been thrilling: the plentiful streams of clear water crisscrossing the valleys, the sweet smell of lodgepole and ponderosa pines. Willows and wildflowers line the creek banks, and groves of aspen turn the world a-flicker with the faintest breeze. If the prairie feels epic, the Black Hills are intimate and cozy.

The hills are especially spectacular in their southeast corner, where the otherwise gentle slopes break apart into cliffs and pinnacles of granite. White settlers dubbed the area "The Needles." Lumps of stone form enigmatic outcroppings—a closed hand, something that looks like a heart, faces in the boulders. These are old, old hills, long under the sky. (Geologists estimate that some formations in the Black Hills are up to two billion years old and have never been covered by ice or ancient seas.) The ground bulges with thick slabs of white quartz and flakes of pearly mica. Sometimes along the trail I came to areas where a mica deposit had split into a million pieces, making the path sparkle in the midday sun. In the Black Hills, the dust literally glitters.

For a plains people, the heights were no doubt stunning. The highest mountain in the range, the 7,200-foot Harney Peak, is the tallest point between the Rockies and the Pyrenees. Traditionally, young Lakota men would go to a "secluded, high place" in the hills to seek visions. Lakota women, too, sometimes experienced visions among the buttes and ridges. It was on Harney Peak that Black Elk had his famous revelations. "The high and lonely center of the earth," he called it, a line that could come straight from Keats or Byron if you didn't know better.

A cave that is always blowing. The ground aglitter. High mountains piercing the sky. Is it any wonder that a place of such marvels was invested with sacred meaning? The first white settlers also caught the spirit of the place: one formation not far from Mount Rushmore is called "Cathedral Spires."

By the 1850s, the steady flow of emigrants along the Oregon Trail had begun to depress game populations as the settlers hunted deer and pronghorn and their cattle herds disturbed bison grounds along the Missouri River. (This was, of course, just a prelude to the bison slaughter to come.) The nations of the Upper Plains understood that the whites would continue to encroach, and that they would probably have to cede some space to them, even if they didn't like it. In 1857, an unprecedented gathering of tribes took place at *Mato Paha* (today known as Bear Butte, outside Sturgis, South Dakota) to formulate a common policy toward the whites. They agreed that they would fight to ensure that the Black Hills—which Lakota leader Red Cloud called "the head chief of the land"— would remain theirs.

In the end, the Indians stayed true to their promise never to surrender the Black Hills. The United States government, however, did not keep its word to leave the Black Hills to the Indians.

In 1866, after a couple of years of steadily raiding wagon trains headed along the Bozeman Trail, Red Cloud led a combined Lakota-Cheyenne-Arapaho force of 2,000 warriors in a stunning victory over the US Army at the battle of Fort Kearny. Almost 100 US troops were killed in what whites referred to as "the Fetterman Massacre." News of the defeat shocked Americans back East—and led to a stunning concession by the US government.

Under the Fort Laramie Treaty of 1868, the United States agreed to Lakota demands for the Powder River Basin and the Black Hills. Some 31 million acres, including the Black Hills, were reserved for "absolute and undisturbed use and occupation" by the Lakota. The Lakota would retain "the right to hunt" on 50 million acres stretching between the North Platte River and the Yellowstone River. Never before (or since) had the United States ceded such a large territory in defeat. The stars and stripes were struck from Fort Phil Kearny and Fort Reno and the outposts were abandoned. Red Cloud's warriors burned them to the ground.

The peace was short-lived. White encroachments into Lakota territory soon resumed—and then increased. In 1874, Civil War hero George Armstrong Custer led a combined military-scientific exploration into the heart of the Black Hills, looking for gold. Custer's dispatches to the East trumpeted "beautiful parks and valleys . . . unlimited supplies of timber" and "gold among the grass." US newspapers inflamed gold-rush fever with headlines like, "Struck It at Last! Rich Mines of Gold and Silver Reported Found by Custer."

The US Army made a pretense of trying to keep prospectors out of the Black Hills, but the government was eager for gold, the financial Panic of 1873 having caused a liquidity crisis. The Lakota and Cheyenne watched as swarms of miners headed into the Black Hills. By the winter of 1875–76, at least 4,000 whites were illegally occupying Paha Sapa.

Conflict was inevitable. Lakota attacks on the miners precipitated a massive, three-pronged US Army offensive against the Indians in the summer of 1876. The campaign did not go as planned. On June 25 of that year, a reckless and arrogant Custer (against the advice of his Crow scouts) rode his force directly into the middle of the one of the largest encampments of Plains Indians ever seen. The Battle of Little Big Horn marked yet another disaster for the US military. By sunset, a combined Lakota-Cheyenne force had destroyed Custer and his entire troop of 210 men.

The Lakota had won a famous battle, but they couldn't win the war. As even the most militant warriors understood, the industrial might of the United States was too great to overcome. In April 1877, Crazy Horse laid down his arms and led 900 people "into the agency"—to life on the reservation. Five months later he was killed, bayoneted by another Lakota who was making a crude attempt to arrest him. The Hunkpapa Lakota warrior-chief Sitting Bull retreated to Canada, but four years later returned to the United States, surrendered, and joined the other Lakota on the reservation. (Sitting Bull was assassinated in 1890, just weeks before the massacre at Wounded Knee.)

In the meantime, the US government seized the Black Hills. In February 1877, even before Crazy Horse surrendered, the US Congress passed a bill unilaterally retaking the contested territory. Even though the Lakota were by then reliant on government rations and were essentially starving—and even though many of them were addled by or bribed with whiskey—only 10 percent of Lakota men living at the agencies signed the treaty ("touched the pen," as the Lakota said) giving up the Black Hills. But the treaty of 1868 clearly stated that three-quarters of Lakota men would have to agree to any future agreements ceding land. As far as the Lakota were concerned, the Black Hills were still theirs, in legal right if not reality.

Much of that sorry story is well known. This next part is less famous, but just as incredible.

Beginning in 1921, a series of indefatigable lawyers working on the Lakota's behalf spent close to sixty years in an effort to press a claim stating that the Black Hills had been taken in violation of US law. The case went through a roller coaster of dismissals and appeals as it bounced among courts and the United States' Indian Claims Commission. In 1974, the commission ruled on the Lakota's behalf and awarded the tribe $17.5 million for the taking of the Black Hills. After a few more years of legal maneuvering, in 1980 the US Supreme Court upheld the Lakota's position by an 8-to-1 decision and awarded the tribe $106 million in damages, a figure based on the value of the gold and silver that had been extracted from the area. Justice Harry Blackmun wrote, "A more ripe and rank case of dishonorable dealings will never, in all probability, be found in our history."

But the Lakota refused to take the settlement money. "*The Black Hills are not for sale,*" Tom Poor Bear told me, thumping the table with each syllable and repeating the line that has become the foundation of modern Lakota nationalism. Today the money remains in an account at the US Treasury. Thanks to compound interest, the sum has increased to more than $1.3 billion.

Did you get that? The poorest, most neglected, worst-treated people in the United States are sitting on more than a billion dollars that they will not touch because it would violate not just their pride, but their religion. During my time on the Pine Ridge Reservation, nearly every person I spoke with repeated some version of a saying ascribed to Crazy Horse: "One does not sell the land one walks on."

The Lakota's 150-year-long refusal to sell the Black Hills reveals something essential about the Native worldview toward wild nature. Yes, the earth is here to be used. And, at the same time, there are some places that hold a numinous power, places where humans are permitted only to visit, and even then only with the holiest intentions. *Sacred* comes from the Latin, "to set apart." Like all peoples, the Lakota held some areas distinct and removed from the rest of the world—a practice that isn't all that different from the modern ideal of wilderness.

"We say all land is sacred," Milo Yellowhair told me. "But some places are *more* sacred."

Harney Peak and Bear's Butte were—and remain—places to fast and to seek visions, places for young men to learn their real names. The same is true of *Mato Tipila*, or Bear's Lodge, the stunning monolith of stone in northeastern Wyoming that settlers dubbed "Devil's Tower." For the Lakota and Cheyenne, Bear's Lodge was a sacred place central to their "star knowledge"—their way of telling time and marking the season. (Devil's Tower was the first national monument established by Teddy Roosevelt under the Antiquities Act, nice proof for how awe crosses cultures.)

At Bear's Butte and Bear's Lodge and the peaks of the Black Hills, the Lakota could explore the mysteries of existence—and mystery is at the heart of Lakota religion. The word *wakan* is often translated as "holy." And so an idea like *Wakan Tanka* is interpreted as the "Great Spirit," a singular God. But, according to Lakota religious scholars, a more accurate interpretation of *wakan* is "mysterious." Wakan Tanka, the animating force of the universe, is the "Great Mystery." To call a thing holy is to deem it incomprehensible.

"We have certain points where we can receive communications from the Almighty," Harold Salway–Left Heron, director of the Oglala Lakota's parks authority, told me. He is a staff keeper and pipe carrier for the Lakota, well studied in the tribe's traditions. "Our people used to go and fast for four days and four nights, and cry for a vision from the Creator, from God, or however you identify holiness. They would receive messages from the other world, from our ancestors that have gone before us. Instructions on how to heal, for predictions, for medicines, for things that we do not have in this physical world. The people would go to that spot continually until they created a portal to another world, a portal to communicate with sacredness."

It turns out that prayer is one more way through which we can create changes in the land. By setting aside some places as sacred, we engage in an interaction with wild nature in which we do not take our sustenance from the earth, but instead make an offering to it. To construct "a portal to another world" through ceremony or ritual or private meditation is to create another type of working landscape—one that is at work by being a sanctuary and a site of communion with the wonders of Earth. Prayer, too, is a use of the landscape. It's how we can give back to wild nature, by doing what humans do best: investing a place with meaning and with myth.

I wish I could end there. It would be nice to think that the Lakota idea of a sacred place as a "portal to another world" could tell us most of what we need to know about how to relate to wild nature today. Black Elk said "anywhere is the center of the world," and I like the ecumenical spirit there—the affirmation that each of us can find sacredness wherever we like, and it doesn't have to be a remote wilderness. But of course it's not so simple. To have such a tidy finish would be dishonest to Lakota history. There is, I think, a darker lesson to be taken from the Lakota experience.

It came to me as I rolled into the Rapid City airport. The sky was perfectly clear, the Black Hills smudging the western horizon. Already I was feeling the sting of nostalgia. I missed the afternoons spent driving in circles around the rez—Porcupine to Wounded Knee, through Manderson, where Crazy Horse's people now live, then up to the pinnacle of Chimney Butte, and back around again. I missed the mornings I spent in the Black Hills, when I woke in my sleeping bag to what the Lakota call *anptaniya*: the breath of day, vapors raised by the sun. *Morning's glory* expressed in a single word, a foreign language shedding new light on a familiar experience. In the rental car parking lot it all seemed so far away.

The modern relationship to the wild, I'm sorry to say, is all too similar to the Lakota's relationship to the Black Hills. The twenty-first-century wilderness is, above all, a place of longing. It's a homeland we are unlikely to able to return to anytime soon.

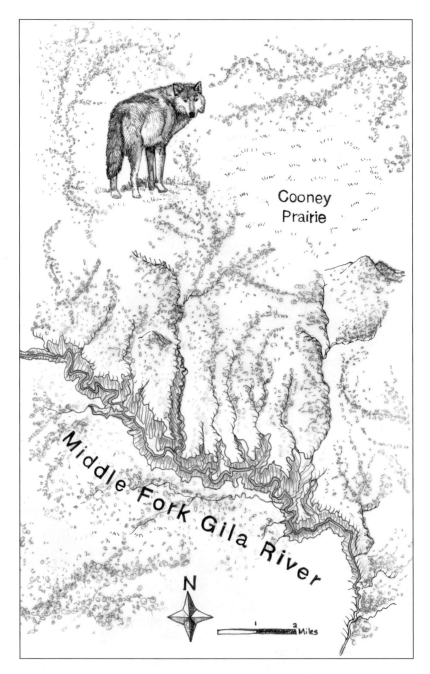

Cooney
Prairie

Middle Fork Gila River

N

1 2 Miles

Gila Wilderness

6

The Ecology of Fear

THE COYOTES WOKE ME UP. They were near, I could tell, almost inside our camp, just beyond where the vehicles were parked. A whole pack, howling in the moonless night—a chorus of coyotes on the edge of the wilderness. The sound was clear in the stillness, the wind having settled after days of bluster. I fumbled around the tent for the travel clock. *1:31 a.m.*

I held my breath to listen better. There came another howl— this one from the other side of our camp, to the northeast. It was much different from the others, a deeper voice, almost baritone: *Aaarrrr-oooooo.* A storybook sound that went beyond memory and into imagination.

A wolf?

I wouldn't let myself believe it. In the weeks leading up to this expedition to New Mexico's Gila Wilderness, I had told myself again and again not to expect encountering any wolves. Forget a needle in a haystack—stumbling across a wolf would be more like finding a single, specific hayseed in a haystack. The best I could reasonably hope for was the imagined presence of the wolf, the thrill of simply knowing it was out there.

These were not the famous wolves of Yellowstone I was tracking. Since they were reintroduced to the Northern Rockies in the mid-1990s, the Yellowstone wolf packs have become celebrities—the focus of Disney documentaries and television specials and countless biologists' studies. Some of the wolves in the Lamar Valley packs have been photographed and filmed to death—literally. (By now the animals are so habituated to being watched by humans that they don't exercise enough wariness when they leave the safety of the national park, and have become an easy mark for hunters.) Instead, I was after the Mexican gray wolf, the Rocky Mountain wolf's smaller cousin. Less famous, perhaps, but just as polarizing: hated, celebrated, beloved, "protected," and "managed" at huge cost and effort.

According to the most recent count at that time, there were eighty-three known wolves in thirteen packs roaming the wooded mountains of Arizona and New Mexico. Just eighty-three animals, and yet they had caused so much conflict among their human neighbors. The ranchers in this vestige of the Old West were outraged about wolves preying on their cattle. The hunters—an important constituency in a place where elk outnumber people—were ticked off about the competition for game. Some moms worried that the wolves would attack their children in a modern replay of *Little Red Riding Hood*. Environmentalists, meanwhile, were angry that the wolf reintroduction was moving so slowly and were worried that the tiny population was at risk of suffering a genetic bottleneck. Caught in the middle was the US Fish and Wildlife Service which, in an effort to placate everyone's fears, had resorted to an elaborate system of control over the wolves.

By the time I arrived in the Gila, the Feds had functioning GPS or radio collars on forty-eight animals—more than half of the known wolves in the wild. Many more had been implanted with PITs (personal ID tags) just like some people put into their pets, with a specific Social Security–like number complete with genetic information and vaccination history. The animal tracking and pack management had become a nonstop job. Every Monday an airplane staffed with personnel from either USFWS or the Arizona Game and Fish Department was spending five to six hours flying over the rugged countryside to pick up the telemetry signals from the radio collars. Whenever a she-wolf had a litter, the wildlife managers would enter the dens to take a census of the pups. The wildlife managers stalked the wolves across the landscape and set out baited, rubber-toothed foot traps in order to nab the animals and get a radio collar around their necks.

In spite of all the manhandling, the wolves continued to live according to what was left of their instincts, which, above all, spur them to roam widely in search of prey, or a mate, or a fresh territory in which to form a pack. When I went to the Gila (pronounced *Hee-la*) in the spring of 2014, the wolves were still confined to the Blue Range Wolf Recovery Area—the government's lingo for where the Mexican gray wolves were "permitted" to be. At 4.4 million acres (about 6,800 square miles), the recovery area was larger than the state of Connecticut. Yet the wolves inevitably ranged beyond it. And when they did, the government agents would swoop in by helicopter, dart and muzzle them, recapture them for a potential re-release back inside the boundaries of the recovery area, or, in the case of "problem wolves," remove them from the wild forever.

A wild animal living in a giant, invisible cage—the whole thing seemed absurd. The Mexican gray wolf's situation all-too-perfectly captured the plight of the animal kingdom in the Age of Man.

From what I understood, the Southwest wolves were mostly wild. They were hunting elk, eating deer, and forming packs. Fifteen years after the first reintroduction, more than 95 percent of them were wild-born and wild-raised. Yet they were held tightly on the leash of our laws. If they crossed some undetectable dotted line—ate one calf too many or roamed too far—they would be put in a kind of conservation jail cell.

1:32 a.m. the clock now read. The howling had stopped as soon as it began. The coyotes had resorted to yipping and yapping and, from what it sounded like, running back and forth, clearly agitated about something. Fearful.

Had that really been a wolf howl? I wondered again. *No way,* I thought.

I tried to close my ears to the coyotes, whose persistent barking was already becoming annoying. I remembered Thoreau: "All good things are wild and free." The Mexican gray wolf seemed to meet that first description. But here on this garden-planet, where room to roam is scarcer than ever, can we really call them "free"?

✑

The cage measures about eight feet by four feet, made from wood and wire and with a corrugated tin roof pitched to the rear. It's well built and airy. There's a door that latches on the inside, benches to sit on, and it's tall enough to stand up in. That's because the cage isn't meant for animals, but rather for human children. The local kids are supposed to go inside the cage when waiting for the school bus, just in case a wolf is stalking them.

There are several such "kids' cages" in Catron County, the sprawling New Mexico county in the midst of the Gila National Forest, and one of them sits at the edge of Heather Hardy's place in the community of Cruzville. Hardy, a single mother of four, is terrified of the wolves. She used to raise laying hens, but then she started to lose them to predators—to a wolf, she is sure. One night in the fall of 2008, she heard commotion among her horses, and then the kids on the porch yelling, "Mom, mom, get your gun." She came out of the house to find a wolf standing on the slope above one of her two corrals. Hardy—a Navy veteran who served as a corpsman in Desert Storm—fired a couple of warning shots into the hill with her snub-nosed .38. The wolf didn't flee. Then she aimed at the animal and fired again. "I got a gut shot on it," Hardy told me. Investigators eventually found bite marks on Hardy's quarter horse, she says. "I don't like to kill anything, but I had to. . . . I just have no patience for those damn things."

Hardy is a sweet woman, with long, brown hair and eyes as green as a wolf's. A tattooed vine wraps around her left wrist. She is, she confesses, an animal lover. Hardy works as an EMT, and neighbors and family bring her injured jackrabbits and squirrels that she nurses and then releases back into the woodlands (except for one squirrel her kids have kept as a house pet). Her juniper-studded lot is like a mini-menagerie: she has two goats, two riding horses, one miniature horse, and seven dogs, plus a small flock of chickens and turkeys. The morning we met, a couple of cats were slinking about. "I love all the wild animals," she said. "I just don't like the ones that they put here, that they raise by hand and then dump on us."

The wolves, she said, are more vicious than other predators. "They kill things—it's a thrill kill. It's more of a game to them. I've seen five calves down, and only one is eaten. My chickens and turkeys, they would kill them but then only take one bite out of them."

There's something wrong with the wolves that have been reintroduced to the Gila, Hardy says. "They are not acting like they are supposed to. They don't have the normal behaviors. Everything is scared but the wolf. I've been hiking and have seen mountain lions—they don't want to get aggressive with you. A wolf doesn't have that sense. They half want to play with you and half want to eat you."

She paused, then said, "Everyone around here is so afraid, because they know what will happen if they shoot one. They know they will go to jail and pay fines out the ying-yang.

"You know, everyone's terrified."

Here are the facts. *Canis lupus baileyi*, the Mexican gray wolf, is a distinct subspecies of the more common *Canis lupus* found in the Northern Rocky Mountains and up into Canada and Alaska. (The timber wolf of northern Minnesota, *Canis lupus occidentalis*, is another subspecies.) While the wolves of the Rockies tend toward dark gray or even full black, the Mexican wolf is lighter colored—more of a yellowish gray, often with a black back and tail tip, and sometimes streaked with auburn. The desert wolf is about a third smaller than its northern cousin. Males usually weigh sixty to eighty pounds and measure roughly five and a half feet from tail to nose, females a bit slighter.

In all other respects, the subspecies are the same. The Mexican gray wolf—often referred to by its Spanish name, *lobo*—is a prodigious traveler. A wolf can travel forty miles in a day, working itself into a "harmonic gait" in which the back paw falls exactly where the front had landed, a rhythmic jog that conserves energy. It has impressive stamina and speed. Wolves have been known to swim for up to fifty miles.

The wolf is, famously, a formidable hunter. Packs will run prey for hours before accelerating to an attack in which they can reach a top speed of thirty-five miles per hour. The animal's anatomy is made for destruction, its forty-two teeth adapted for seizing, tearing, and crushing. Its jaw can slam shut with 1,200 pounds-per-square-inch of force, twice the power of a large dog, enough to snap a bone. In the Southwest, lobos' preferred game are ungulates—elk, mule deer, white-tailed deer—supplemented by rabbits, mice, squirrels, and, if they can catch them, birds.

Wolves also prey on domesticated livestock: sheep when they find them, and the occasional horse, but mostly cattle—especially pregnant cows and young, vulnerable calves. Usually they hunt in packs. The wolves come from behind, snapping at the prey's haunches and hams and biting at the flanks until the animal weakens or is brought down by a lunge to the throat, thrown to the ground to be ripped open and, typically, eaten alive.

Sometimes wolves will engage in what biologists call "surplus killing"—a killing spree in which they will hunt more than they can eat. Perhaps this is driven by the wolf's "search image," a picture of a specific prey that is burned into the animal's mind in its adolescence; having learned to take down one kind of prey, it will keep coming back to that food source. Perhaps such surplus killings have to do with the fact that a wolf's life is marked by feast and famine. Wolves have been documented eating up to a fifth of their body weight at one time; other times they go weeks without eating anything. A wolf can spend up to a third of its life on the hunt.

Wolves are sophisticated social animals that live in complex communities. A lobo pack normally consists of six to eight animals, dominated by an alpha male and alpha female that are the best hunters in the group and (usually, but not always) the only ones that breed. Gestation is exactly sixty-three days, during which the alpha female digs a den for raising the pups. She has her litter in the spring, normally between four and six pups, a third of which won't make it to adulthood. The pups are weaned at five weeks, at which point the whole pack works together to feed and care for the young.

Wolves are social eaters as well as social hunters, and biologists have speculated that this cooperation is what creates interlocking and overlapping bonds of responsibility and accountability among the pack—a primitive system of ethics, if you will. Packs also develop distinct personalities, so that an experienced field biologist can tell one pack from another just by its behavior. A wolf pack creates a culture unique to itself.

The animals are smart as hell. Wolves have a highly developed "response intelligence"—that is, they learn. A wolf that has been trapped once is all but impossible to trap again. Wolves have been found to defecate on human artifacts—beer cans, spent ammunition casings—as well as poisoned meat, as if in a kind of warning to other wolves.

Above all, they are indefatigable. In his landmark book, *Of Wolves and Men*, Barry Lopez shares the story of an aerial hunter who, in the winter of 1976, encountered ten gray wolves traveling on a ridge in the Alaska Range. The animals had no way to escape, and the gunner killed nine quickly. As Lopez tells it:

> The tenth had broken for the tip of a spur running off the ridge. The hunter knew the spur ended at an abrupt vertical drop of about 300 feet and he followed, curious to see what the wolf would do. Without hesitation the wolf sailed off the spur, fell 300 feet into a snowbank, and came up running in an explosion of powder.

Live free or die, I guess. For more than 200 years that normally meant death. Like most other wolves in the Lower 48 (the timber wolf being the one exception), the Mexican gray wolf was nearly hunted into oblivion. Lobos were trapped, hounded, and gunned down from the air. Thousands were poisoned by strychnine-laced meat scattered from airplanes, like a chemical air raid raining death from the sky. It is no exaggeration to say that the nineteenth- and twentieth-century campaign against wolves—spurred by ranchers and often funded by the state and federal governments—was an attempt at biological genocide. Extinction was the goal.

(The war against wolves and the wars against the Indians overlapped and were all but undistinguishable. At the start of an 1865 campaign against the Northern Plains tribes, a US Army general told his troops that the Cheyenne and Lakota "must be hunted like wolves.")

By 1976, there were only a handful of Mexican gray wolves remaining and the animal was placed on the endangered species list. Roy McBride, a longtime wolf hunter whom the federal government had hired to rescue the last ones, trapped four males and one female in the mountains of northern Mexico and brought them to the United States. Two more wolves came from a line of animals descended from a pup that a Canadian tourist brought across the Mexican border in his motorcycle saddlebags and decided to drop off at the Arizona Desert Museum in Tucson. Biologists had to re-create a healthy population starting from just those seven animals.

Captive breeding facilities in California, New York, and Missouri crossbred the animals in order to establish some genetic diversity in the population. By the late 1990s, the captive population was around 170 animals, and the USFWS—after conducting scores of public meetings across Arizona and New Mexico—said it was time to begin releases into the wild.

The first release, in the spring of 1998, went badly. Of the eleven wolves reintroduced to the wild, five were shot and killed almost immediately, three were removed and returned to captivity after leaving the recovery zone, and one went missing. But the Feds decided to keep going. In November of that year, Interior Secretary Bruce Babbitt—a former Arizona governor—participated in the release of two female wolves and declared at a press conference that

> The public wants the wolves. These are public lands, a part of every schoolchild's heritage. And this is how we treat them? . . . Cattle growers think they are entitled to produce the maximum

possible number of cattle they can ship to the stockyards every fall, and they believe they are entitled to do this on public lands regardless of what the public wants from these lands. . . . The reason we're releasing these two wolves is to send a message that this is public land. . . . The Mexican gray wolf has come home and it's here to stay.

Bold promises, to be sure. But the situation on the ground was more complicated. Constantly hammered by the region's livestock industry (and likely hamstrung by political appointees during the George W. Bush administration), the USFWS spent most of the following decade in a defensive crouch, and the recovery effort stalled. After an initial burst of activity, releases essentially stopped. Between 2004 and 2013, only eleven more wolves were sent into the wild. Meanwhile, the USFWS and Arizona Game and Fish staffers were busy either "translocating" wolves—meaning capturing, penning, and then re-releasing—or permanently removing them. Between 1998 and 2013, 104 wolves were translocated and another 24 were permanently removed from the wild. At least eight wolves died in the course of such operations.

About half of those translocations or removals were prompted by livestock depredations and about a third were due to wolves going beyond the recovery boundaries. Eight percent of USFWS management actions involved "lethal control"—that is, government agents shot and killed twelve wolves. And that's a fraction of the wolves that have been killed illegally. Since 1998, at least fifty-five wolves in the Southwest have been shot in violation of federal law. Poaching is the number-one cause of death for the lobo.

After hitting a nadir of forty-two animals in 2009, the population was at eighty-three wild wolves by the time of my visit, though everyone on all sides agrees that there are more lobos out there, un-collared and unknown. (By February 2015, the number of confirmed lobos in the wild had risen to 109 individuals in nineteen packs.) Some 260 wolves remain in the captive breeding facilities or at Ted Turner's vast Ladder Ranch in south-central New Mexico. Take that in: there are more than twice as many Mexican gray wolves living behind wire fences as there are in the wild.

While many other wolf populations in the United States have been removed from the endangered species list due to intense lobbying from the livestock and hunting industries, the lobo continues to receive Endangered Species Act protection. Under a new USFWS rule finalized in January 2015, the wolf recovery area was expanded to the east and south, allowing lobos to roam all the way to the Mexican border and opening the way for captive wolves to be introduced into New Mexico (before that, initial reintroductions only occurred in Arizona). But the wolves still will not be permitted to pass north of Interstate 40 (which connects Flagstaff, Arizona, to Albuquerque), and the wild population will be capped at 325 animals. Private individuals will be given wider latitude to shoot or harass "problem wolves."

This compromise approach has pleased no one. Ranchers and hunters, who are already opposed to the reintroduction program, say the range expansion will only lead to more attacks on cattle and game. Conservation groups say the 325-animal figure isn't close to what's needed for a healthy population, and they argue that the range increase should connect with the Southern Rockies of Colorado. The Mexican gray wolf may have been given a somewhat larger box in which to roam—but it's still a box.

Those are the facts. Everything else about the lobo is hearsay, slander, exaggeration, fabrication, aspiration, or plain old myth. We seemingly can't help but drape animals with our symbolism. It's too bad, really. Most often we end up smothering the plain eloquence of the thing-in-itself under a pile of metaphors. But maybe that's just a natural (I guess you could say) part of human instinct. As a species, we're hardwired to seek out emblems through which to interpret the world. "We use wild animals to tell stories about ourselves," Jon Mooallem writes in *Wild Ones*, his book about the "psychic pack animals" we moderns have devised. I can't think of a beast that has been asked to carry more of our psychic burden than the wolf.

⸙

Animals are a portal into wildness. With their autonomy and their native indifference to us, wild animals force us to consider that other beings have a will of their own, a set of interests distinct from ours. This is especially true for us twenty-first-century city slickers who have grown unaccustomed to anything beyond our ken. Just the glimpse of an animal in the wild—the flash of fur in the underbrush, a tail bounding out of sight—is like an otherworldly visitation.

I suppose I could have chosen another beast through which to explore the ironies and idiocies of our relationships to wild animals. The Yellowstone bison, say, brought back from the very edge of extinction after the wanton slaughter of the 1800s, and now hemmed into the national park, hazed or gunned down if they roam too far because ranchers are afraid that the buffalo might make their cattle sick with brucellosis. Or the California condor, another remarkable recovery story: a stable population bred from just a handful of birds, then reintroduced to the Ventana Wilderness of the Big Sur Coast. Now back in the wild, the condors are at risk from lead poisoning (bullets and buckshot being a dangerous dietary supplement for a scavenger). The mountain lion also would have been a good choice. The big cats are making a precarious comeback after a century of bounty hunting—slyly moving from mountain strongholds to eek out a living in the Hollywood Hills, the suburbs of Denver and Salt Lake City.

But I kept coming back to the wolf, blessed and cursed with charisma. Of all the large carnivores in North America, no other animal provokes such intense feelings of both attraction and repulsion. For some people, the wolf has long been the totem animal of wildness. Adolph Murie (Olaus's brother), a longtime ranger in Alaska's Denali National Park and a pioneer of carnivore biology, once said, "Wolves are the voice of wilderness." For others, wolves are a hated menace. That great outdoorsman, Teddy Roosevelt, scorned the wolf as "the beast of waste and desolation." Centuries earlier, European Christians feared the wolf as the devil in disguise. The Cheyenne and the Arapaho viewed the wolf as the spirit-animal of power and courage, but the Navajo believed it was a witch. A New Mexico rancher, Joe Bill Nunn, told me, "These animals are terrible animals, these wolves. They are brutal killers, they are savage killers. There are no two ways about it."

Our conflicting emotions about the wolf, it appears to me, have more to do with our species' similarities than with any differences. We're more like wolves—with their big appetites and their guile—than we are like the naïf-ish deer. Read a bit of *canis lupus* biology, and after a while the wolf tales begin to sound Shakespearean—a tumult of rapaciousness, generosity, fratricide, outcasts and loners, loyalty and affection.

I feel sorry for how the wolf has been freighted with our parables, at this point a weight heavier than any radio collar. The wolf, I make myself remember, is just an animal, not any more "solvable" than human nature is solvable. Toward the end of *Of Wolves and Men*, Barry Lopez, after having spent hundreds of pages plumbing the depths of our mixed-up wolf myths, delivers a stern warning against overthinking: "We assume that the animal is entirely comprehensible. It seems to me that this is a sure way to miss the animal and to see, instead, only another reflection of our own ideas."

Okay, then. Once we strip away all of the legends, what do we have left? The plain fact of brutal competition. If some hunter-gatherers esteemed the wolf, agriculturalists have always despised it—and for some good reasons. Our relationship with the wolf is so vexed because, perhaps more than any other animal, the wolf is our direct competitor. For millennia, a wolf pack at the edge of the pasture meant the difference between a season of bounty and a season of famine. The wolf snatches dinner off the table.

This ancient rivalry forces hard choices: Can we find a way to live with the wolf's wildness and share space together? Can we coexist, and come to see another carnivore as something of an equal, and not just an enemy? Or do we have to control it, and in that control limit its wildness, the very thing that draws us to it?

The Gila is big country. At 3.3 million acres, the Gila National Forest is one of the largest US Forest Service holdings in the Lower 48. Combine that with the adjacent Apache National Forest, Cibola National Forest, the Bureau of Land Management's Plains of San Agustin, the Blue Ridge Wilderness in Arizona, and the White

Mountain and San Carlos Apache Reservations, and you get a vast area of some 6 million acres of wildlands. That's a space about as large as New Hampshire, with a combined population of fewer than 50,000 people. When driving the region's back roads, the locals bring along an extra can of fuel and four or five gallons of water. If you break down, you could easily have a fifty-mile walk before finding help.

The Greater Gila is mostly high desert, beginning around 4,000 feet and rising to peaks of 11,000 feet or more. The terrain is furrowed and creased, with the usual extremities found in arid lands. Mountains tumble into ridges, which fall into mesas, which slide into ravines and mazes of canyons. Most of the area is a mixed woodland of pinyon and juniper stitched with yucca and prickly pear. At the higher reaches there are stunning, miles-long ponderosa groves. All around grows what's called grama grass—a bunch grass with a filament stalk topped by a sickle-moon seed head. In the fall, after the monsoon rains sweep through, the grass is blue-green. Then it turns brown-to-blond. For most of the year the Gila looks like a pure-gold carpet dotted with the green bulbs of the juniper-pinyon stands.

The Gila is sometimes referred to as "the Yellowstone of the Southwest," and the big, mostly uninterrupted space is an ideal wildlife haven. On the Plains of San Agustin I've seen herds of pronghorn bounding across the great sea of grass beneath a rain-shredded sky. Tens of thousands of elk and deer graze the woodlands. There are javelina, fox, coati, and bighorn sheep. Beaver are still found along the Gila River, where needlelike spires rise above sycamore stands. In the springtime the woodlands are filled with pinyon jays and mountain bluebirds. When the sun dips past the last ridges, the canyon tree frogs begin their all-night trilling and a flawless firmament appears. I've never seen such stars: the night so pure the sky has its own texture and the stars appear hung in three dimensions, crystals dropped into a net of dark matter.

The Gila Wilderness is one of the last places in the continental United States big enough and wild enough where even someone with a good bit of woodcraft can go astray. Few people make it out that far, and the trails are barely used. Many of the paths are little more than faded cow tracks or game trails that peter out into underbrush. Stop, listen, and an awesome stillness arrives, as if someone took the Quiet Dial and turned it all the way to zero. There's just red rock and gold grama grass and the twisted shapes of dogged juniper. For a desert kid like me, the Gila is heaven on Earth.

The Apache clearly thought so, and it's no wonder they fought so hard for the place. Geronimo himself—né Goyathlay, "One Who Yawns"—was born and raised at the headwaters of the Gila, and I'm sure he learned some of his squirreliness in that canyon labyrinth. Before the Bedonkohe Apache splashed in the hot springs of the Middle Fork of the Gila, the Mogollon lived in the region for centuries. A great many of them, or so everyone agrees. According to what a hunting outfitter in the hamlet of Luna told me, some archaeologists believe that from Alpine, Arizona, to Glenwood, New Mexico, as many as 10,000 Mogollon might have lived along the San Francisco River, a pit house dug into every rise. I can believe it. Despite the dryness (about fifteen inches of rain falls in a "normal" year), the land has a richness about it—precisely the right mix of grass, trees, and game.

Of course, the Indians' heaven was also the cowboys' idea of paradise. The green-and-gold grama grass is just as suited to cattle as it is to elk, and the open vistas and the parklike spacing of the juniper and pinyon make the Gila easy terrain for herding longhorns or traveling cross-country by horse. It's the very ideal of high desert rangeland, ripped from the pages of a Louis L'Amour Western. Such natural wealth was an obvious point of conflict. The Apache Wars raged on and off for nearly forty years until, in 1886, Geronimo surrendered, many of the Apache were shipped off to Florida or Oklahoma, and the ranchers and loggers took over.

Because of its size, wildness, and relative absence of humans, the Gila was a natural choice for the reintroduction of the Mexican gray wolf. There was also a kind of poetic justice at work: the Greater Gila had been the site of one of the most famous conversion tales in the science of ecology.

When he was a young man just starting his career with the US Forest Service, Aldo Leopold was posted to the Gila. It was there— among the "crumpled topography" where "the Creator . . . piled the hills 'high, wide, and handsome'"—that Leopold's commitment to wilderness preservation was formed. The mesas and mountains of the Gila were his model for wilderness as an area "big enough to absorb a two-week pack trip."

No doubt the land's lonesomeness influenced the notion of wilderness as a place empty of people. By the time Leopold arrived, only echoes of memory remained of the Bedonkohe Apache. In 1924, after years of internal lobbying, Leopold got the Forest Service to declare the Gila a roadless "primitive area"—the United States' first official wilderness preserve, created forty years before the passage of the Wilderness Act. (There's an important irony in the fact that a permanent human settlement—a Mogollon-era village, now Gila Cliff Dwellings National Monument—sits at the center of "America's First Wilderness.")

The Gila also forged Leopold's insights into ecology, especially the role of predators in an ecosystem. Leopold was a man of his time, and like other land stewards in the early twentieth century, he believed predators were a scourge to be eliminated. "I thought that because fewer wolves meant more deer, that no wolves would mean a hunters' paradise," he wrote in *A Sand County Almanac*. One day he and some other foresters were "eating lunch on a high rimrock" when the party spotted a she-wolf and pups frolicking on the banks of a river below. "In those days we never heard of passing up a chance to kill a wolf," he wrote, and "in a second we were pumping lead into the pack." Leopold climbed down to the carnage and reached the mother wolf "in time to watch a fierce green fire dying in her eyes."

Later, after years of studying how ecosystems work, Leopold would recognize what a mistake it had been. Without wolves, the deer population exploded, the deer began to eat too much, and the woodlands started to suffer. In his short essay, "Thinking Like a Mountain," Leopold wrote that a landscape without wolves "looks as if someone had given God a new pruning shears, and forbidden Him all other exercise. . . . I now suspect that just as a deer herd lives in mortal fear of its wolves, so does a mountain live in mortal fear of its deer."

Contrary to all the myths and legends he had grown up with, Leopold concluded that predators also have a place in nature's design. Yet he knew that such a truth would be hard for many to hear: "Only a mountain has lived long enough to listen objectively to the howl of a wolf."

The night that I heard the coyotes and what, I hoped, were wolf howls, I was camped at a well-used hunter's camp at a spot called Cooney Prairie, right on the north edge of the Gila Wilderness. It was my second night in the Gila wildlands as a small group of wildlife advocates and I tried to locate some of the lobos.

Peter and Jean Ossorio had organized the trip and picked the site for our base camp. The couple is in their early seventies, and they've dedicated their retirement to wolf advocacy, especially Jean, whose enthusiasm for the lobo is unflagging. Since 1999 she has spent more than 350 nights in a tent in the wolf recovery zone. During that time she has spotted a wolf forty-three times, and has many more reports of tracks, scat, and howls in the woods. Jean says her passion for wolves developed in the 1970s, when she had a chance to see up close one of the lobos in the captive breeding program at the Endangered Wolf Center outside St. Louis. "Part of the reason I was interested was that they are very social animals, their relations in the pack are very interesting," she told me. "Part of it was that they were so maligned and vilified. I have always identified with the underdog, or the not-quite-so-popular." She keeps up a blog, *Lobos of the Southwest*, that sends our regular e-mail action-alerts to wolf advocates. When it came time to celebrate her seventieth birthday, she took an overnight trip to wolf country.

Ossorio's fervor has kept her youthful. There are only a few strands of gray in her long brown hair, and when she talks about the wolf—which she does in an encyclopedic, looping stream-of-consciousness—she's seized with a girlish enthusiasm that seems incongruent with her Bea Arthur–like build. "The more you see them, the more you become fascinated with their regulatory effect on ecosystems," she says. On her wrist she wears a silver bracelet stamped "Makas 131," the name of the alpha male of the original Hawk's Nest Pack, the first wolf she ever heard howl in the wild.

Her husband, Peter, served twenty-eight years as an artillery offi-
cer in the army, including several tours in Vietnam, then went to law
school on the GI Bill and became a federal prosecutor. Eventually
Jean infected him with her wildlife passion. He is, as it were, a patriot
for the wolves. Mention the wolf controversy, and his normally easy
demeanor snaps into the cold bearing of an experienced litigator.
"The ranchers are grazing their cattle on public lands. That's a privilege, not a right, and it comes with certain responsibilities," he said.
"I don't care what the ranchers think. I care whether they are com-
plying with the law of the land."

We were also met at Cooney Prairie by Dave Parsons, a former
US Fish and Wildlife biologist who launched the wolf recovery program in the Southwest. Parsons ran the reintroduction effort from
1990 to 1999. Then, he says, "I defied direct orders to cook the science, and it led to my early departure from the agency." According
to Parsons—whose arguments are backed by several peer-reviewed
studies, including a draft recovery plan created by the USFWS in
2011 but never adopted—a healthy, stable lobo population would
mean about 750 wild animals distributed across three distinct population segments. Parsons told me, "The three regions would be the
Blue Range Mountains, the current recovery zone; the Grand Canyon eco-region, into Southern Utah; and the Southern Rockies,
meaning southern Colorado and Northern New Mexico. In that
750 scenario, the density of wolves in a given area is very small.
That's what the best science calls for."

Accompanying Parsons was an English journalist, Adam Nicholson, on assignment for the British literary journal *Granta*, which
was planning an issue themed—get this—"Wild America." After
one evening at the camp with the five of us, we were joined the
next morning by Michael Robinson, a campaigner with the Center for Biological Diversity. Robinson is a cherubic fellow whose
frequent smile and gentleness belie a fierce passion for wildness. I
had met him the previous autumn, during a USFWS wolf hearing in Albuquerque, and he had made no attempt to hide his dis-
dain for the ranchers. "If you're trying to increase biodiversity, using
cattle as a tool makes about as much sense as creating peace with
a machine gun," he told me. "Part of the culture of the livestock

industry—despite its image of rugged individualism—is about asking for government assistance." The ranchers, he said, act like they're "rural royalty."

Later that day Craig Miller from Defenders of Wildlife arrived. He came rolling up on a dust-covered, red BMW cross-country motorcycle, rumbled through the camp, leaned the bike against a juniper, and promptly cracked open a beer. Miller has devoted close to twenty years of his life to the wolf reintroduction, much of it spent driving the deep back roads of the Greater Gila to meet with livestock owners (the motorcycle had been an in-kind gift from someone who wanted to see the work continue). Defenders of Wildlife has been at the forefront of wolf recovery efforts in the United States, mostly trying to work with ranchers to ease their concerns about the reappearance of an apex predator. The group started by paying ranchers direct compensation for confirmed wolf depredations of livestock, then helped pay for range riders to follow the herds. "We're trying to get the ranches to go from coexistence to tolerance to acceptance," Miller said, an epic weariness apparent in his voice.

We spent that afternoon sitting on the edge of Cooney Prairie, watching two large elk herds graze and talking about the wolf saga. Everyone there had a profound, almost ineffable love for the wolf. Yet there was something else at work: a hope that in bringing back the wolf, some larger wound would be healed.

The lobo recovery in New Mexico and Arizona is just one part of a broader movement for "rewilding." All too aware that the preservation of wild places is no longer sufficient, conservationists have turned a lot of their attention to ecosystem *restoration*. We must also repair the damage of the past: freeing the rivers that have been impounded, reviving degraded wetlands, nursing endangered species. If twentieth-century conservation was about drawing lines on a map, twenty-first-century conservation is about filling them in.

The enthusiasm for ecological restoration—especially rewilding, with its emphasis on the return of large predators—marks an important turning point for the broader environmental movement. Rewilding flips conservation from a defensive, rear-guard action into a forward-looking act of imagination determined to create more abundance. Rewilding affirms that we don't always have to play the role of destroyer. Our interventions with wild nature can be virtuous, too.

All kinds of ecosystem restoration efforts are under way across the United States. Many of them are small-scale—the repair of this single streambed, the rebuilding of ecological processes in that one watershed. To restore ecosystems on a regional or continental scale requires, above all, the reappearance of predators.

There's a scientific term for the phenomenon of ecological knock-on effects that Leopold observed during his wolf-hunting days in the Gila: "trophic cascades." Mostly using the Yellowstone wolves as a test case, wildlife biologists have confirmed that apex predators like wolves exert a profound pull on an ecosystem. This has to do with what is known as "top-down regulation." An ecosystem is shaped and reshaped in several ways. "Bottom-up regulation" refers to the flow of energy coming up from the great mass of fungi and bacteria and the photosynthesis of plants, consumed then by thousands of different invertebrate and vertebrate herbivores. Top-down regulation is the way in which animals at the top of the food web mold an ecosystem. Apex predators influence the behavior of their prey, and that new prey behavior in turn affects the species on a lower trophic level. The mere presence of a top carnivore ripples through the landscape.

Imagine: a wolf appears on the scene. Suddenly, the elk can no longer loaf around the valley bottoms. They actually have to start paying attention to their surroundings and looking for threats. As the elk become more cautious, they begin to browse differently. Trees and shrubs are offered a reprieve. Aspens, once chewed to the ground, reappear along the riverbanks. The more robust greenery offers new space for other critters. Beavers come back. Mesocarnivores like coyotes begin to behave more cautiously. Cause-and-affect spills from one level of the food web to another, like a waterfall. The mark of the wolf's tooth, biologist Cristina Eisenberg says, is powerful enough to shape the course of a river.

Biologists sometimes describe apex predators' influence on the landscape as "the ecology of fear." The phrase intuitively makes sense. An elk herd without wolves nearby enjoys the luxury of becoming stupid and lazy. As soon as a predator returns to the picture, the elk have to become alert and active. Fear invigorates them. The elk's new skittishness gives an aspen—and riverine grass and, eventually, a beaver—more of a chance to thrive. The presence (or absence) of apex predators is the most important single predictor of how wild a landscape can be.

And so the wolf once again becomes a symbol greater than itself: the animal as optimism for our ability to rebuild the wild world. Rewilding, Craig Miller said, represents nothing less than an "evolution of ethics." The centuries-long war against wolves, he said, "was all part of this campaign of 'winning the West,' this struggle of man versus nature. Well, we won the West. It's ours. We own it. The question is, having won it, do we have to beat it into submission? Or can we strike a balance, because our health and our welfare as a society ultimately depend on the persistence of wild nature."

Robinson said, "People today have a sense of how badly out of whack our world is, and here's an animal that's so vital to restoring that balance. We have to start somewhere, and the wolf is a great place to start. This is an animal that can be instrumental to conserving a large ecosystem. . . . It's an issue of justice, of making things right."

I shared their enthusiasm. But I worried that the rewilding effort had gone astray—not in the intent, but in the execution, which seemed at once half-assed and heavy-handed. After fifteen years there were just a scant eighty-odd wolves in the wild, nearly all of them micro-managed. The lobo program seemed like a bad example of gardening on a landscape scale.

We decided to go for a short hike to get a better view of the elk herds. The thought had been that the elk would be an attractant to wolves, but there were no signs of the predator. We spotted some very old, almost petrified scat, and that was it. Looking for a wolf in the wilderness? It seemed a fool's errand.

Miller and I were walking out ahead of the others, and I asked what motivated him to keep going despite the serial setbacks. "The reason I'm into this—why I do it, personally—it's because I think the wolf is an amazing way for us to reinterpret our relationship to wild nature and to each other. It triggers something in us, both the good and the bad. The wolf starts important conversations. I was at a meeting with a rancher recently, and he stopped and looked at me and said, 'Why are you here?' And I told him, 'I'm here because I feel deeply about wild places and wild *life*.' And that's what it is. The wolf makes us think about how we want to relate to nature."

Laura and Matt Schneberger's place is just about the prettiest ranch you've ever seen. From the closest neighbor's house to Rafter Spear Ranch it's a fourteen-mile drive down rough and tumble Forest Service roads. Then you hit a perennial creek that cuts a narrow valley between ponderosa slopes and rugged cliff faces. Big, tall cottonwoods shade green pastures. The solar-powered ranchhouse, bunkhouse, barn, and saddle shop are all painted a rustic red, the same color as the old Farmall tractor in the yard. Horses and pack mules mill about the corral. Laura keeps beehives and a vegetable garden out back by the windmill. It's like the place was conjured from a modern cowboy fantasy.

The Schnebergers, Laura especially, have distinguished themselves as two of the most vociferous opponents of the lobo recovery program. Laura is one of the hubs of a network of ranchers and hunting outfitters who have fought the Feds' program (and the wolves themselves) for more than fifteen years. She's like a mirror of Jean Ossorio. Mother of three, grandmother, with the unflagging energy of a pioneer. She's the longtime president of the Gila Livestock Growers Association (Matt's the vice president), and has turned the group's website into a clearinghouse of anti-wolf news. She writes frequently about wolf depredations, keeps up an e-mail list for area ranchers, and never misses a public hearing on the animal, which she abhors for what she says it has done to her herds. "A wolf needs twenty pounds of meat at every sitting, or forty pounds, depending upon the type of wolf," she told me. "Once they learn to kill your baby calves, they'll kill a baby calf every day. They'll clean it up, and you won't have anything left."

On my way to meet up with the Ossorios and Dave Parsons, I spent several days driving around the back roads and meeting with ranchers and outfitters to try to understand their hatred of the wolves. What Craig Miller and Michael Robinson see as an act of restorative justice, the ranchers view as an imposition by "animal lovers" and "bunny huggers," a vicious plot by the federal government and well-heeled environmentalists to destroy the rural way of life. "Anybody else would be allowed to protect their property, but apparently if you ranch in wolf country, you are not," Laura Schneberger said at one point in our hour-long conversation. "Your rights are different than everybody else's."

The antipathy toward wolves begins with an understandable fear. Wolves eat cattle. For ranchers, that fact represents a real economic cost as well as an emotional burden. There's the death of marketable head and the financial loss from undersized animals that, due to stress from being stalked by wolves, haven't put on as much weight as they otherwise would. Then there's the psychic pain. Losing an animal that you've raised from the day it was born is a real blow to the heart that isn't covered by the compensation money the Feds or Defenders of Wildlife pay for confirmed depredations.

"We love animals, and we like to take care of our animals and provide safety and nutrition for them, and when we have these killers, these wolves, released, it's sickening to see our animals killed," rancher Joe Bill Nunn told me. "It's devastating on us. There was a reason they got rid of those wolves in the first place."

Wink Crigler, a widow whose family has raised cattle outside of Greer, Arizona, for more than a century, said to me: "In the beginning I believed there could be some coexistence, so that I could exist and there could be some wolves here, too, recognizing that these are public lands with 'multiple use.' Now I don't think there can be any coexistence. Because what I raise is what wolves like. I can't produce that commodity in the presence of wolves."

Then something weird happens. The reasonable concerns about wolves' impact on livestock get magnified and the fears become deeper, an echo of the old myths about the wolf as a devil, a fiend. Many people who live in the Gila are convinced that the wolves pose a threat to human life. Everyone has a frightening wolf story to tell.

"Did you hear about what happened to the Nelson boy?" a woman who ranches to the northeast of the Schnebergers (and who asked to remain anonymous) said to me. "He was cornered by a pack of five wolves. They pushed him back against a tree, surrounded him. He had a rifle, but he was afraid to use it. He said he was afraid that if he shot and killed a wolf his dad would lose his grazing allotment."

Laura Schneberger told me: "My daughter was on horseback, and two wolves held her up. We had several incidents where kids were followed home from the school bus by these animals."

When Wink Crigler heard I was planning on going backpacking into the Gila alone, without a firearm, she tried to warn me off. "You're nuts," she said. "Why do you want to put yourself at risk like that?"

The fears then get magnified to a stronger power. Many residents of the Gila are convinced that the wolf reintroduction is a government conspiracy to wreck the livestock industry and drive people off the land. In the course of my conversations I heard whispered warnings about "Agenda 21"—a United Nations plot to corral people into cities, where they'll be easier to manipulate. The wolf, I was told, was just the vehicle of a larger agenda to crush people's freedom.

"It's like having a four-legged Al-Qaeda around—it's about instilling fear," a local hunting outfitter, Brandon Gaudelli, told me one morning over coffee. "They have allowed in the wolves to get rid of the elk, so that someday there won't be anything left to eat. They want everybody out of the mountains. They want us in the cities where they can control us."

Crigler was certain of the same: "What I really think is the wolf issue really is not that much about the wolf. What's it's really about is Agenda 21, putting people off of the land, taking away the ability to be sustainable. The wolf is a tool to accomplish what the government has been talked into by a lot of environmentalists who . . . share this mentality of moving people out of the rural areas and back into the urban areas."

Such talk is part and parcel of the long-running "Sagebrush Rebellion"—the fear and loathing many rural Westerners feel toward a federal government that they say is out of touch with their needs. The wolf hatred is identity-based: wolves kill cattle, and so ranchers kill wolves, and that's how it's always been. The fear of wolves also springs from what environmental historian Roderick Nash calls "the wilderness condition."

Or, as I like to think of it, "the pioneer's paradox." The frontiersman has a tortured relationship with the wilderness and a deep ambivalence toward wildness. The pioneer loves the frontier—it's the anchor of his self-identity as the kind of person who can make it in an unforgiving landscape. Yet the pioneer has no patience for the Romantic's sentimental view of nature. Life on the frontier is a war of all-against-all; you have kill to survive. And so, almost inevitably, the pioneer through his domination of the land ends up destroying the very thing he values: the freedom of a wild place.

The couple of thousand people who make their home in the Gila do so because they love it there—the big lonesome of open country, the stars at night. "This the wild, wild West," Gaudelli said. "This is the last vestige outside of Alaska. Wildness is freedom, literally. It gives you the opportunity to go out and feed yourself, whether that's hunting or fishing or trapping or raising cows."

Within five minutes of us meeting, Heather Hardy, the mom with the kids' cage in front of her property, told me, "I just like the wild life."

As for Laura Schneberger, who can see the north boundary of the Aldo Leopold Wilderness from her front porch—she wouldn't live anywhere else. Her eyes welled up as she talked about the deep human need to feel connected to nonhuman nature. "It's just a good place to be," she told me. "You're not just around everything that man has made." *And* she is convinced that making a living in such a place requires taking control of it. To thrive in the wild, you have to show the other carnivores who's boss.

"When you're not controlling a major predator—a top-of-the-food-chain, I-can-kill-anything-I-want-to predator—then you're not doing things that would dissuade it from bothering people," she told me. Because of the lobo's endangered species status, "when you have a wolf kill, you cannot just go in there and take [the wolf] and mitigate the problem. We control everything else. We can shoot a lion. We can shoot a bear. We need some kind of control with this animal."

The summer before I visited the Schneberger ranch, the family had been having a problem with a she-wolf attacking cattle in one of the pastures on their property. "1108 came down here, and was killing cattle right in my field," Laura said, referring to the wolf's ID number. "The cows had bite marks on them. One little old female wolf, fighting two cows and calves—because that's what they do."

Laura's husband, Matt, took down one of the rifles the family keeps over their front door, went out, and shot the she-wolf.

The tracks were as clear as the morning-gold light over Cooney Prairie: canine pads, stamped into the red dust of a Forest Service road. Jean Ossorio, wildlife geek, had a tracker's measuring stick. She laid it on the ground next to the marks. Four-and-a-half inches from tip to toe. Too big for a coyote. The storybook howl in the night really had been a wolf.

We spent the next half-hour doing a bit of backcountry sleuthing. The spacing of the tracks seemed off, Parsons and Jean agreed. It was a weird gait, as if the animal had been injured. We followed the tracks down the road, and after about a hundred yards the riddle resolved when a second line of tracks appeared. There had been a pair of wolves, at first trotting single file, and then splitting off to jog side by side.

We were all amazed. A pair of wolves a mere thirty-five yards from where our tents were set! Parsons said we should all buy lottery tickets when we got back to civilization—our luck was that good. Jean bubbled with delight. "I think I can image the thrill a hunter feels when he finally spots his game," she said to me as we walked up the slope to the camp. "It's just so exciting, to know the wolves are close."

The next question was which wolves they were. As I made breakfast, Parsons and the Ossorios pored over Gila topo maps and the most recent radio-collar telemetry reports, taken just days before. There was a possibility that the pair was the Canyon Creek Pack— F1246 and M1252—but they would've had to cover a good bit of ground to reach Cooney Prairie and, besides, it seemed they were getting ready for denning. Perhaps the pair had been some members of either the Dark Canyon Pack or the Fox Mountain Pack, but, again, the wolves would need to be moving far and fast to get to us. With no radio-collared wolves in the immediate area, there was a chance the pair was anonymous. We couldn't be sure, however, until the next telemetry report came out.

We had our breakfast in the sun, maps spread out, speculating about the wolves' whereabouts. Everyone wanted the pair to be uncollared, to be "outside the program," as Parsons said. It was dispiriting to think that even our wild animals are locked firmly in the matrix, their movements as carefully tracked as those of any person with a credit card and a laptop.

None of the wolf advocates like the micromanagement. "Just as the [federal] agencies have boxed the wolves in, they've boxed in themselves—and then they complain that they're in a box!" Michael Robinson said. "The solution would have been, in the first place, to put a lot of wolves in the backcountry and create a viable population. But these people, as an agency, are addicted to control."

Yet with such a small wolf population, the heavy-handed management appeared the unavoidable price of recovery. "At this point in time, we may have no choice," Robinson said. "The management we're doing now is just remedial, just to keep things alive." The tracking is sometimes in the wolves' interest, he said. When a wolf is killed illegally, at least the Feds know where to find the body.

It seemed that, in our efforts to control the situation, we've managed to slip the lobo loose from its symbolism. There was little doubt that the animals were living according to their instincts, but they weren't exactly self-willed and sovereign. If the wolf was the totem of wildness—well then, wildness appeared to have been diluted into insignificance.

We draw Cartesian lines on a map and expect that wildness will abide by the rules. The wilderness goes here; the working landscape goes there; the wolves will remain in this invisible box. It's little more than self-flattery. The legal boundary of a wilderness designation can keep earthmovers out, but it can't keep a wolf in. The Sisyphean effort to restrict the lobos to a certain zone—a task about as practical as tacking olive oil to a wall—says a lot about the limitations of our legal wilderness system. A wilderness area might be able to preserve some of the world's wildness, yet wilderness can't contain the wild. That's the thing about wildness: even in its debased condition, it sneaks by.

The wolf war in the Southwest is fueled, above all, by a clash of instincts. The wolf's instinct is to roam far and wide. Our instinct is to dominate, to shape the world to fit our needs. We also have a deep desire for omniscience. More than any other animal, the wolf tests our ability to live with things out of our control and beyond our understanding. Unlike much of the rest of wild nature, the wolf isn't merely indifferent to humans and our desires—the wolf is actually *antagonistic* to our interests. As Laura Schneberger put it: "We're both apex predators, and so we are in direct competition with them."

At its heart, the fear of the wolf is a fear of wildness, and the fear of wildness is the fear of a loss of control.

"They should be controlled, they should be hunted," Brandon Gaudelli said. "I don't know if there's a place for the wolf anywhere out there anymore. Maybe if you had a huge fence you could control them, and have people come out and watch them."

Rancher Joe Bill Nunn said, "If you want to see a wolf, and enjoy seeing a wolf, and want to show your kids a wolf, the place for that is the zoo."

Kerrie Romero, a representative with a hunting organization called New Mexico Outfitters and Guides, seemed to capture the thinking of most locals when I heard her say, "We spend thousands of hours in the backcountry. We understand the importance of a healthy predator–prey balance in the wild. We also understand that in a world of human authority, that dynamic needs to be managed in order to maintain balance."

A world of human authority. At stake in the wolf rewilding is whether we have the grace, sometimes and in some places, to forgo using that authority. The ancient struggle between wolf and man forces us to question what—if anything—we are willing to sacrifice to accommodate the needs of other beings. The wolf makes us ask whether we're willing to share space on this planet.

The question is made more difficult by the diminishing size of our world. With 7 billion people on the planet, space is at a premium. Wink Crigler says there might once have been a place for the lobo. But not today. "It isn't wild anymore," she said to me, waving her hand at the hills of her ranchland. "This isn't 1800. The wilderness that people envision in their minds—it's gone. This place might seem wild to you, but it's not wild when ten miles away there's a town, and twenty miles away there's a Walmart. Therefore it's not wilderness for the wolf anymore, either. I think it's a bad thing."

I think it's a bad thing, too. But I disagree with her conclusion. While the reality of the Anthropocene complicates the task of rewilding, it doesn't make it impossible.

Rewilding is a gift of forbearance, measured out in human patience and a generosity toward other creatures. There's no question that, as a nation, we're wealthy enough to afford such generosity, which amounts to little more than the cost of some cattle. The problem is that the costs fall entirely on a small number of people who bear a disproportionate burden. Justice for the wolves means injustice for some people. Which is precisely why it feels so hard.

If we really want the wilderness to remain wild, it will require that we find some way for grace to overcome instinct, some way to cultivate a selflessness to which, as a species, we are untrained. After all, it's easy to love a nature that just looks pretty. It's an entirely different task to live with a nature that is threatening—quite literally, the wolf at the door.

Three days after we heard the howl at Cooney Prairie, Adam Nicholson, the English journalist with *Granta*, got to join the USFWS and Arizona Game and Fish agents on their weekly Gila flyover to

track the wolves' radio signals. Parsons and the Ossorios had carefully briefed Nicholson on how to handle the situation: he wasn't
supposed to say anything about the tracks we had spotted, and
instead would play dumb. He would casually ask the government
officials if they could point out to him Cooney Prairie—*just out of
curiosity, mate*—and then listen on the sly for any wolves in the area.
If the wolves were in fact renegades, we wanted to keep it that way.

Nicholson spent a good part of a morning crisscrossing the Gila
from the air, and as soon as the plane landed he shared his report
with Parsons, who called Jean Ossorio, who e-mailed me and the
others. The plane hadn't picked up any radio collars close to Cooney
Prairie. The nearest collared wolf pair was the Canyon Creek Pack,
some twenty miles to the northwest and now dug in for spring denning. Jean concluded: "I would be very surprised if this pack came
over to Cooney Prairie to howl and make tracks on the morning
of 3/28 and then trotted back over to T Bar Canyon in time for the
3/31 flight. It's possible, but not all that likely."

The wolf pair we had heard was off the grid and outside the
matrix.

The news seemed a minor miracle. I flashed on the ending
to *Jurassic Park*—engineered animals finding a way to procreate.
I thought of how life, in all of its unruliness, defeats any enclosure. I
thought again about the limits of lines of a map, about the resilience
of mystery. Somehow, against all odds, the wolves howling in the
night had been wild and free.

Sawtooth Ridge

7

Back to the Stone Age

THE FIRST TIME I SAW LYNX VILDEN, I thought I had slipped
into George R. R. Martin's *A Game of Thrones* saga and had
found myself north of the Wall, in the land of the wildlings. This was
a few years back—August 2012—and my partner, Nell, and I were
backpacking on the eastern side of Washington's Cascade Mountains,
among the so-called Golden Lakes of the Sawtooth Ridge. During
the previous two days we had encountered only one other person,
a motorbiker who startled us one morning as he tore through the
trails of the national forest. Besides that, the place was ours. So it was
a surprise that evening when—as we played cards near the edge of
Sunrise Lake—Nell grabbed my arm, hissed "*Look over there,*" and
pointed to a lone figure moving along the lakeshore.

In the dusk light it was hard to tell whether it was man or woman.
The getup didn't help. The person was clothed all in buckskin:

buckskin breeches and a buckskin jacket with a fur-trimmed hood pulled over the head. A long, thin knife in a leather sheath hung off a leather belt. A fishing pole was in his—or her—right hand. The figure moved quickly and stealthily through the trees. And then it was gone, disappeared into the shadows. Around the fire that night, Nell and I played the scene over and over. Who could it be? Were there other deerskin-clad strangers out there in the woods? But we saw no other signs of people—no campfire in the distance, no voices in the dark.

The next morning, after a somewhat restless sleep, the mystery partly resolved itself. While I made breakfast, Nell decided to scout around the lake, on the pretense that she was off for a bowel movement. I was having my first sip of coffee in the glow of the early sun when I saw her running down the slopes of purple lupine. As she reached our camp she said, "They're making *traps*."

Nell had crept through the larch groves until she came to a shelf of rock tucked into the mountainside and thought, *If I lived up here, this is where I would be.* Right then, coming over a small rise, she saw a man sitting next to a fire circle. His back was toward her, but she could see that he was hunched over some kind of assemblage of wood and fiber, crafting or repairing it. He was wearing a big black vest that looked to be made of bear fur. He had long, black dreadlocks.

Nell was telling me all of this when two people appeared down at the water's edge: the person from the evening before with the fur-trimmed hood, now clearly a woman, and another woman, dressed in a buckskin blouse and buckskin skirt. I had to learn more and so I headed over to talk with them.

I've always found the etiquette of the backcountry to be a little tricky. To be sure, you want to be friendly to strangers on the trail. At the same time, few people go into the wilderness for chitchat, and it's best to give folks their space. A plain "Good Morning" is usually enough, and that's what I said as I strolled up to the pair.

The woman we had spotted at lakeshore at dusk seemed like she had spent a lot of time in the wild. Her corn-silk hair was pulled back into thin, blonde braids, and there were deep creases around her brilliant, blue eyes. She was sharp-featured, with a hawk's nose, and the skin on her hands and face was weathered and tanned. She had on a pair of well-worn hide sandals. She looked tough.

She asked me my name and I told her. "I'm Lynx," she said. The plot thickened: her voice had an unmistakable English accent.

We made small talk. Lynx asked how long we were out for. Just three days, I said. I asked how long they were out for. Three weeks, she answered. The whole time, the other woman just stared at me coldly. She was barefoot, her feet as hard and sooty as charcoal, and I got the sense that my mere presence was offensive. I found myself suddenly self-conscious about my modern gear—my polyester long johns and camp sandals, my Patagonia outershell and wool cap. I felt kind of like a faker.

Curiosity was killing me, but I didn't want to be too nosey, so I wished them good luck. The friendly Lynx said the same, and I went back to the cup of coffee waiting for me at our tent site.

Not more than ten minutes later we saw the trio getting ready to leave. The man in the bearskin vest and the barefoot woman were wearing on their backs large wooden baskets stuffed with buffalo-skin bedrolls. Lynx had a small fur-and-leather pack no bigger than a breadbox. There couldn't be much more in there than some dried meat, I recall thinking.

Just before they tramped off down the trail, Lynx came over to talk with Nell. I was washing the breakfast dishes down at the water, and missed the exchange. I came back to camp and asked what they had talked about. Nell said I wouldn't believe it. They wanted batteries—*AA batteries*. They needed them for their digital camera.

"The Stone Age Living Project" is the name for the scene Nell and I had stumbled upon. Lynx Vilden wasn't that hard to track down. Her Internet tail is long, including a pretty nice personal website, www. lynxvilden.com. She's something of a rock star in what is sometimes referred to as "the ancestral skills" community. She travels around the world—from ancient cave sites in the south of France to the deserts of Africa—conducting trainings in primitive technologies. Her Paleolithic repertoire includes bow making and trap construction, hunting, animal processing, hide tanning, wild-plant foraging, making tools out of stone and bone, and fire starting, among many other skills and crafts.

She has been at this a long time. Lynx was born in London in 1965, a time when, as she would tell me later, "being a little girl in England meant staying clean and looking pretty." Her instincts pointed in a different direction—toward "going out and playing in the woods and getting dirty." She divided her time between her father and stepmother, who were salespeople for a company that made plastic cutlery, and her mother and stepfather, who were artists. (Her stepfather was principal of the prestigious Saint Martin's School of Art.) Lynx took the artist's path. She went to theater school, where she majored in choreography. She dove into the punk scene, and spent three years in Amsterdam partying hard because "what the hell, the world was going to explode anyway."

Eventually she realized that she was "killing myself faster than I wanted to." During a visit to her grandmother's place in Sweden her life changed. Amid the wildness of the Nordic woods she realized that she didn't need to drink and use drugs. "The forest saved my life," she says.

Lynx (I never asked for her given name) came to the States, where she fell in love with the American wilderness and with an American man, with whom she had a daughter. She was drawn irresistibly to the wildest places. But, as she told me, she was "lazy, really," and didn't like carrying all of the weight required for a long trek into the backcountry. She figured she could lighten her load if she could find her own food, and she began studying hunting and wild foraging. In 1990 she made her first fire using a bow drill. "I found my passion at the tender age of twenty-four—and that's my story," she told me.

She has lived in wildlands ever since. For close to a decade she and her daughter lived in New Mexico, in the Taos area. Their place was an eight-mile hike from the nearest road. Eventually her daughter (who had grown up with squirrel-skin finger puppets for toys) said she wanted a "normal life" and moved to Seattle to be with her father. Lynx followed her to Washington and found a place alongside the Twisp River, right at the edge of the vast Okanogan-Wenatchee National Forest on the eastern slope of the Cascade Mountains.

Today, Lynx's small patch of woods is the home of the Living Wild School, whose motto is "We'd rather *live* in the wild than *survive* in civilization." The property serves as the base camp where she and others spend a season getting ready for their weeks-long, late-summer trip into the wilderness equipped only with Stone Age clothes, tools, and foods. "The Project" is how Lynx and the other wildlings refer to their experiment in fully primitive living. "After four months preparing, we go time-traveling," Lynx likes to say. That bizarre morning on Sunrise Lake, Nell and I had happened upon the eighth annual Project.

In case you missed it, the primitive is having a comeback moment. Many people swear by the "Paleo diet"—which means lots of meat, nuts, and fruits, and no processed grains or oil—supposedly what our hunter-gatherer ancestors would have eaten. There's a *Paleo Magazine* dedicated to "modern-day primal living." Cable TV is chock-full of survivalist "reality shows." Surf the channels and you'll find *Survivorman, Dual Survival, Naked and Afraid, Live Free or Die*, and *Ultimate Survival Alaska*. There's even a show on the Weather Channel titled *Fat Guys in the Woods*. Lynx told me that a month doesn't go by without her receiving a pitch from a TV producer to do some kind of survival show. She refuses because "it's not anything about cooperation, which is what we need. It's a competitive thing that they're putting out."

There is, of course, a long modern tradition of fetishizing the primitive. It started with the Rousseauian celebration of the Noble Savage, continued through James Fennimore Cooper's Leather-stocking Tales, and reached something of a climax as Teddy Roosevelt celebrated "the strenuous life." At the height of the early-twentieth-century primitivist fad, a guy named Joseph Knowles became a national celebrity after venturing nearly naked into the Maine woods and living off the land for months. His dispatches from the wild—written in charcoal on pieces of birch bark—were a media sensation. (Knowles was later accused of being a fraud.)

We moderns have some deep desire to reconnect to the raw and the primal. We yearn for the wild. We want to be assured that it still exists, and we want to experience it—even if that means vicariously through our TV sets. The fascination with the primitive is evidence of how we are all suffering from what environmental journalist George Monbiot calls "ecological boredom." Numbed by the human landscapes of the suburb and the city, we're hungry for some taste of a more visceral, more intense way of living. We want to be reminded that there's a wilder world somewhere out there.

At first, the Stone Age Living Project's radical atavism just seemed like the thirst for the wild taken to the extreme. I thought of it as the backcountry version of Civil War reenactments—weekend warriors playacting as cavepeople in the woods. *Who are these people?* I wondered. And what are they hoping to accomplish?

As I learned more, I discovered that Lynx was part of something much bigger and much deeper. Across the United States there is a loose network of primitivists dedicated to keeping alive some glimmer of the old ways. It's an underground movement of modern-day nomads and hunter-gatherers. I heard of a couple—Moira and Ray—who wander the national forests and BLM lands with a herd of goats that provide them with meat, milk, and portage. I learned of a similar nomadic goat herder, Cannon, who moves from the Arizona high desert to the low desert with the rhythm of the seasons, consuming more than a gallon of goat milk a day as he goes. I was told stories about Finisia (Fin) Medrano, who follows old Shoshone routes through the Great Basin by horseback and covered wagon. From New Mexico to Washington to California, I kept hearing tales of a guy named "Barefoot Doug" who had spent years living in the wilderness, much of it—yes, you guessed it—buck naked.

There are regular meetings of such folks, a "gathering of the tribes," if you will, when the primitivists come together to trade skills and hone their ancestral crafts. Rabbit Stick in Idaho is the oldest. There's an annual camp outside of Phoenix called Winter Count, and an annual spring convergence in California called the Buckeye Gathering. Down the road from Lynx's place occurs a fall gathering, the Saskatoon Circle.

Primal living, Lynx and others believe, is a way in which to encourage an ethic of ecological responsibility. "We've got to get to know the earth again," Lynx would tell me when, in a private conservation, I asked what, exactly, she was hoping to accomplish. "We can't really get to know her if we're always behind walls and beneath roofs and above floors, you know. I think if people could actually take off their shoes and feel the earth, sleep on the ground and feel the energy, be outside and feel the weather, eat from her—just to get to know her. It's kind of like, imagine if you get to know a new person, you're much more likely to take care of and protect that person."

Creating that kind emotional connection is more likely to be long-lasting, Lynx believes, if it's unmediated by modern technology. "I love it when people get that sense of awe and wonder. When you become a part of your environment, and if that environment is wild, then *you* become wild." Rediscovering the oldest lifeways is a path through which to resolve the ancient tensions between wild nature and human culture and technology. Going primal is a way to prove that people aren't separate from nature. "Why can't we think we're *of* the wild?" Lynx said. "Trying to get our niche back in the web of connectivity—that's something to strive for."

There are, of course, some human societies that have managed to remain (out of sheer isolation or out of choice) more or less in the Stone Age: hunter-gatherer tribes in the depths of the Amazon rainforest or the hinterlands of Papua New Guinea. Lynx's Stone Age Living Project represents something unprecedented—a *return* to the Paleolithic past from the Anthropocene present. Against all the currents of history, Lynx and the wildlings are attempting to explore whether it's possible for modern people to capture something from the primeval, bring it back to the present, and see if it can change our modern frame of mind. *Epochnauts*, I guess you could call them.

The endeavor is every bit as ambitious—and every bit as vexed—as the campaign to reintroduce wolves to the Gila. It's nothing less than an attempt at *human rewilding*.

❦

When I contacted Lynx the spring after running into her in the mountains, I was hoping that I would be able to accompany her on one of the Stone Age excursions. Lynx quickly disabused me of the fantasy. If I wanted to go on "the Project," she informed me via e-mail, I would have to participate in her four-month-long immersion course. That's the time it takes to make your own clothes and tools and gather and dry your own food, all of it Paleolithic style.

Unfortunately, I didn't have time for that kind of commitment. I would have to settle for Lynx's Basic Skills Class, a one-week crash course in the elementals of primitive living that she teaches at her property, located about ten miles upriver from a town called Twisp. And so, on a September day, I found myself driving through the dramatic mountain passes of the North Cascades in a beat-up GMC Suburban that had been modified to run on veggie oil. The beast of a vehicle—nicknamed "Falcor"—was owned by a primitive skills enthusiast, Jamie, who had just finished a stint teaching at Alderleaf Wilderness College, a survival school not far from Seattle. She had a handmade bow in the back and a deer-hunting license she was eager to use. Her gray-green Carhartt pants were as much patches as original fabric. Before heading over the mountains, Jamie and I picked up a few other classmates. Craig and Sherie were a couple from Australia, where Craig teaches something called "natural movement" that combines martial arts, tai chi, and gymnastics. Also in our carpool was Shauna, a part-time yoga teacher and part-time gardener from Vancouver, British Columbia.

The other course participants would turn out to be just as eclectic. Charles was a long-haired troubadour (clarinetist, classical guitar player) from coastal Quebec who had hitchhiked his way across the continent to get to Twisp. Another Quebecois, Katherine (or Kat), had heard about Living Wild from a fellow employee at the natural-foods store in Ottawa where she works. There was a French-Spanish guy from Lyon—Eddyr—who was getting ready to spend the year with Lynx in preparation for the 2015 Stone Age Project. (Thanks to a 2013 documentary that aired on French TV, Lynx is big in the Francophone world.) A fifty-something fellow named Sylvan was also getting ready to spend a year at Living Wild; in an earlier chapter of his life, he had spent fourteen years at an ashram in Virginia. Rounding out the group was Willa, a twenty-one-year-old from New York City who skipped college to pursue an education in ancestral skills and who had recently completed a year at a survival school in Wisconsin called Teaching Drum. She had lost her bag on Greyhound, but she came well equipped with loads of crazy stories about surviving off of squirrel, raccoon, and opossum. "I can always tell someone who has done Teaching Drum," Lynx said one night over supper. "They're always hungry."

Lynx's place was a sight. Her scant three-acre parcel was scattered with all sorts of primitive shelters in various states of construction. Here, a hand-built lean-to made of logs and pine needles. There, a tent cabin and a yurt. Many of the shelters consisted of little more than a tarp slung on a line between trees and anchored by shipping pallets. Skins hung from tree branches. Hides were stretched over logs.

As we arrived, we were greeted by some of the long-timers who had spent the summer (or, in a few cases, longer) living with and learning from Lynx and then going on the 2014 Project, which they had completed only a week earlier. A guy named Matt welcomed us. He was wearing the kind of tight, thigh-length shorts popular among hipsters in Oakland and Brooklyn—only they were made of deerskin. He had on a headband and a necklace with a black stone hung on a sinew cord, and nothing else. With his almond-shaped eyes, coppery skin, and chiseled chest, he was like a Pleistocene Adonis.

The women were just as attractive. I'll admit, I had expected the women would be hairy and scary. But here were these rosy-cheeked and bright-eyed women, fresh-faced lasses from Ontario and Devonshire who had traveled halfway across the planet to explore a primal way of life. And in the middle of it all was Lynx. She was like a primitivist Pied Piper, a den mother in deerskins whose positive spirit (none of that Doomer and Prepper stuff for her) had attracted a clan of idealistic twenty- and thirty-somethings.

The low-impact, high-tech style of wilderness adventuring that involves lots of fancy gear has a reputation for being so expensive that it's exclusive. I figured I would find much the same with this particular wilderness subculture; after all, it takes some measure of privilege to go into the woods for months on end to prepare for going totally Stone Age. But while some of those at Lynx's camp did come from affluent families, most of them were from humble backgrounds. Long-timers Jane and Jessie had been working at a nature camp outside of Exeter, England, when they dropped everything for the Stone Age Living Project. Jane had been the camp cook, and it didn't sound like the pair of friends came from a posh circle. Another long-timer, Emma, had been living at her parents' house in Ottawa while working at a grocery (she was the one who spread the word to Kat). Eddyr had been involved in the *Indignado* protests of unemployed youth in Spain. He and Charles both seemed to be living hand-to-mouth.

Regardless of background, all of the wildlings and aspiring wildlings shared a certain alienation with the postmodern world and a gnawing curiosity about life before humans domesticated much of Earth. That first night we sat around the fire passing the "talking stick" from person to person and sharing our stories. At one point, one of the Englishwomen, brown-haired and green-eyed Jessie, said, "I really liked the idea that I would have nothing separating me from nature—no Gore-Tex, no metal knife. Those things are useful, but they're also a buffer. I wanted a sense of immersion, to be more immersed than I've ever felt in the wild, or with nature. And I wanted that immersion with a tribe, or a clan. Going on that journey with other people was really important to me."

Jessie said the experience of living entirely Stone Age had allowed her to see afresh civilization's imprint on the natural world. It was as if a veil had been removed from her eyes. "Our impact was so much less heavy than it would be in the normal world, but it was so much more visible than it would be in our daily lives," she said. "After a few days, we had beat a path between the shelter and the fire circle. Normally, you don't see where the oil is mined. That was what it was for me—recognizing the impact we have and getting more comfortable with that. Living wild—it gives you a real, direct sense of what it means to be human."

In the firelight, the stars bright and clear above, Lynx shared an entry from her 2014 Project journal (paper, pencil, and corrective eyewear being some of the few "modern exceptions" allowed). The passage had to do with the psychological vertigo some people experience while living fully Stone Age. Lynx read to us: "This land, this world away from the madness, brings up so much questioning and deep introspection. We act like society's refugees. We are in shock, suddenly unable to function in this new land. Have patience. It will come."

We were standing in a circle on a beach of smooth rocks at a bend in the Twisp River. Lynx was wearing buckskin breeches rolled at the cuffs and a fringed buckskin tunic that showed off her rope-muscled arms. There were curved, inch-long pieces of carved bone in her ears, and a beautifully wrought stone knife in a sheath hanging from her neck. She was barefoot.

"The great thing about learning from the earth is that we can be in communion with everything around us," Lynx said. "Start looking around, and you probably see a bunch of rocks. But I see a mountain of treasure. Every one is special and beautiful and unique. They are precious for all different reasons. I barely know where to start. All of them, separate beings. There's *fire* in some of these rocks—you can smell it."

When she's teaching, Lynx's theater background is apparent. She's a natural performer, with an actor's hungry joy when hitting her lines—"*a mountain of treasure*," as though she were reciting Robert Louis Stevenson. It was Day One of the Basic Skills Course. Our assignment was to gather the materials for a bow drill from what we could find on the river bend.

A bow drill is a tool for making fire—basically, a stick that is spun against a flat piece of wood until the friction creates enough heat to ignite some tinder. The tool consists of a *bow* into whose *string* you twist the *spindle*, aka the drill. The wooden drill spins against another piece of wood, a flat one called the *baseboard* that sits on the ground. To keep the spindle straight you use a handhold called the *socket*, usually made from bone or stone. By sawing the bow back and forth you can create enough heat between the spindle and the baseboard to "birth a coal."

Lynx suggested pliable willow for the bow. A piece of alder would be a serviceable baseboard. Red-osier dogwood makes an excellent drill. Before sawing off a demonstration willow whip with her stone knife, Lynx bowed slightly to the plant and said, "Excuse me." When she had her bow piece in hand, she said "Thank you" to the willow and turned with a nod, her perfect London manners matched to an old-fashioned Native American courtesy.

For a socket she suggested one of the thousands of rocks scattered along the riverbank. Soon the ten of us were spread out on the beach, each bent over a pile of rocks. We spent the next hour or more tapping, pounding, knocking, scraping, and sanding our rocks to hollow out a thumb-sized divot in which to turn our spindles. To the rifle-toting deer hunters drinking beer just downriver, it must have sounded like a Stone Age sweatshop.

We hiked the short mile back to Lynx's place, through the pasture where she grazes her three horses. Once back at the main camp, she showed us how to splice deer sinew to make the string for the bow, and we finished the day making cordage. As I twisted the helix of the cord—the sinew literally wrapping itself around itself—I thought of Milo Yellowhair, how he told me that all things work in cycles, circles within circles.

Day Two opened with storytelling by Lynx. Striking a thespian pose, she got down on one knee in a fire-starting position and as she sawed away on her bow drill she started to tell us the story of "Rainbow Crow," a myth from the Chelan people (a mountain Salish tribe) about how humans came to tame fire. She began: "A long time ago, when the world was first new . . . it was the humans that suffered, because they had no clothes yet, they had no fur, and they would get cold. Well, back in those days, the animals and the birds didn't fight with the humans, they actually had sympathy." And so the animals and the birds who saw the People suffering from cold tried to help the People by bringing them fire.

First, Eagle, so big and strong, tried to fly to the sun and return with an ember for the People. "And he flew and he flew until his feathers started to smoke, and he got tired, and he came back and said, 'It's too far. I'm sorry. I can't bring fire back to the People.'" And as Lynx said this, her spindle popped out of her baseboard. Next, Hummingbird, so very fast, volunteered to make an attempt for the sun. "And she flew, and disappeared quite soon, she was so small, and her little feathers started to smoke, but she didn't turn back, she carried on going, closer and closer toward the sun, and finally she, too, became too tired, and came back down." As Hummingbird failed, Lynx's spindle came flying out from the string just as the baseboard began to smolder. Finally, a bird with the most amazing plumage and voice stepped forward—it was Rainbow Crow. She flew to the sun through the heat and fire. "She didn't stop. She carried on going toward the sun, until she was able to reach the sun, and take a tiny little spark from the sun, and she brought it in her beak back down to the People." But the success involved a sacrifice: Rainbow Crow's feathers were singed black and her once-beautiful voice had turned into a raspy *croak.*

As Rainbow Crow succeeded in the mission to the sun, Lynx's coal caught. She carefully dropped the red nugget into a nest of cedar bark, where it smoked, and then, with a few puffs of breath, caught into orange flame.

Next it was our turn. While Lynx had made fire-starting look effortless, I quickly learned that, for a novice, it was anything except easy. I tried again and again, but couldn't get it to work. My spindle wouldn't stay in the socket, and I had to go back and tap out a deeper hole. When I did get the drill to spin, I lost energy before I could get the smoking wood to form a coal. Soon enough the notch in my alder baseboard was deep and polished black, yet I still didn't have a fire.

My classmates were having an easier time of it. Charles quickly birthed a coal. Then the two Aussies nailed it. Soon after lunch Kat and Shauna both turned coals into fire. But as the afternoon wore on I was still sawing away, and beginning to feel like something of a Paleo failure.

Sometime around my fiftieth unsuccessful attempt, it hit me: this was one of my wildest adventures ever, and it was by far the most Promethean.

A celebration of human ingenuity, I was coming to understand, is at the center of Lynx's efforts. She is awed by our ability to reshape the immediate world around us. "*Homo sapiens*—what does it mean?" she said the first day on the beach. "Man the toolmaker. We don't have claws or teeth, so we had to develop something to have an advantage. And that advantage is tool making."

Embracing the essential human gift of inventiveness is at the core of the Living Wild experience. It's not a school for wilderness survival—it's an attempt at *cultural revival*, the long-lost culture of living close to the land. And, even more radically, it's an effort to illustrate that, for most of human existence, wild nature and human culture didn't feel so separate, for the simple reason that our earliest technologies were birthed straight from the raw earth. By returning to the Paleolithic, Lynx wants to show, we can resolve the Edenic rupture that split humanity from the wild. We can become native with nature once more.

The erasure of the boundary between technology and wild nature was one of the most head-spinning elements of the Project, according to Lynx's long-timers. Surrounded by only stone and bone and wood, it was difficult to tell where human invention ended and where "Nature" began. As Jessie said, "Just the aesthetic of looking at my friends and not seeing anything unnatural—there was nothing to take away from the experience." Matt told me: "I felt like I was much more part of the land." This merger between self and wilderness occurred—not by "leaving no trace," as the modern backpacker code of ethics goes—but instead by using intelligence to refashion the elements. To live wild meant embracing human creativity: the sheer awesomeness of stripping tools out of stone and coaxing fire from wood.

The wildlings were just as much in love with their handmade creations as a gearhead who adores her GPS gizmos. Our second night at the fire, Jessie talked passionately about her relationship with the tools and clothes she had created. Tanning, she said, "is like alchemy. You take a dirty, smelly hide, surrounded by flies, and you turn it into something beautiful." While she said this she fingered the hem of her charcoal-stained buckskin skirt as if it were the finest silk.

Lynx's efforts at cultural restoration go beyond fire starting and knife making. She is just as interested in the ancient human arts of community. Every night after dinner we gathered for songs or stories, either around the fire where we cooked and ate or in the "Fire Lodge," a simple but handsome, cedar-shingled octagon pavilion. Lynx uses the fire circle like a portal to the past, to see what we can retrieve from a time when face-to-face contact over a flame was the most important mode of communication.

That second night the women from Devonshire taught us a traditional English folk tune. It was a lilting melody, reminiscent of the rolling landscape of the British Isles. Together we sang:

Home, I'm going home

I need the land to feed my soul

Take me home

Take me home

Over the green, green hills

And far away

Later, as I went to sleep with the song wedged firmly in my mind, I thought once more of the Lakota, and of longing, and of how our home on the land now feels "far away." Although we may strive to return to a homeland in wild nature, it's easier said than done.

The point was reinforced the next night—the third of the course—as I stayed up late talking around the fire with Matt, Shauna, Kat, and Jessie. Matt was telling us about the challenges of the Project, about how far we are from the knowledge of the deep past. "No matter where you go, you're still you," he said. "We can go into the wild wearing skins, but we're still us, with all of our modern baggage."

The Project had tested, in a very harsh way, the wildlings' willingness to cast off that baggage, especially our society's instinct for control. It had been a summer of brutal fires in the Eastern Cascades, as fires tore through 265,000 acres of forest in the surrounding hills, destroying nearly 300 homes. Lynx and her crew had been preparing to set off into the wilderness just as the fires were at their most intense, and some members of the clan were unsure whether it made sense to go. "My civilized mind was freaked out: *This is ridiculous. This is not safe*," Jessie said she remembered thinking. "The smoke was all around, and we didn't know where the fires were. The disasters were conspiring to test how committed we were to going out."

For much of their time in the wilderness, the group was surrounded by great plumes of smoke rising up the shoulders of the mountains. The days were filled with haze, the evenings edged with an apocalyptic glow. Their lungs began to hurt. It felt, Lynx said, like being "an island in a sea of smoke while a lower world burned."

Sitting around the campfire telling us about all of this, Jessie said, "Rewilding humans is much more difficult than I thought. Because my whole brain, my whole mind, is domesticated. . . . We still have what Jane and I called 'Brain Radio.' We were wearing buckskin out in the woods, but Brittany Spears was pumping in my head."

Everyone drifted to their shelters or tents, but I stayed up, watching the flames. I was about to head to bed myself when Willa came running into the firelight. Awkward Willa, with her mop of hair and her big, square glasses, her once-white oxford shirt dusted with so much dirt and ash that it almost matched her khaki pants. We got to talking, and I asked about her story. She told me she was raised by vegetarians in Carroll Gardens, Brooklyn. She said she had often felt disconnected and alienated when she was growing up. Then in high school she had discovered Radiohead, and it made her feel less alone.

I had noticed she had a beautiful singing voice, and I asked if she knew any Radiohead songs well enough to sing. I'm a big Radiohead fan myself, and I suggested "Fake Plastic Trees" from *The Bends* or "Optimistic" from *Kid A*. She picked "The Gloaming," a track from *Hail to the Thief*.

It's a dark, electronic dirge, a postmodern epitaph for the waning hours of industrial civilization. In its frightful way, it felt like a more appropriate campfire song for this Human Age than some old folk tune.

I began to tap out a rhythm on my water bottle. With the yellow light of the flames flickering on her face, Willa sang alone in a low, haunting voice. The song, in part, goes like this:

Genie let out of the bottle

It is now the witching hour

Genie let out of the bottle

It is now the witching hour

. . .

They will suck you down to the other side

They will suck you down to the other side

They will suck you down to the other side

They will suck you down to the other side

To the shadows blue and red, shadows blue and red

Your alarm bells, your alarm bells

Shadows blue and red, shadows blue and red

Your alarm bell, your alarm bell

They should be ringing

They should be ringing

They should be ringing

They should be ringing

. . .

This is the gloaming.

"We learned to manipulate the Earth," Lynx said the first day on the banks of the Twisp River, "and we learned to manipulate the world to such a degree that we're on the verge of making it uninhabitable for ourselves and for other creatures. And it all started with sticks and stones."

Or, in brief: our fires have gotten away from us. Nearly every technology is a double-edged sword that comes equipped with benefits as well as risks. The most obvious example is, of course, the twisted connection between energy and climate change. Abundant energy is a modern miracle, the very lifeblood of our society. And yet, as we know, when we figured out how to burn coal and dig up oil we ignited what is now a global conflagration. We are, indeed as Lynx said, "man the toolmaker." But sometimes those tools can backfire.

The wild is supposed to be a refuge from such worries. The legally designated wilderness is an attempt to keep some places free from the dominance of human technologies. As I've noted, it's an imperfect arrangement: a wilderness boundary can't keep a wolf in, nor contain a wildfire, nor hold global warming at bay. But wilderness can still keep an engine *out*. If wilderness remains significant at all, it's because of the bright line that says, *No motor shall pass here.* In the wild, if nowhere else, the size of space still matters. By forcing us to negotiate the land by horseback or on foot, the wilderness restores distance and scale. On the trail, a mile is made meaningful once more.

Now, however, this core function of wilderness is at risk from some of our newest inventions. The awesome telecommunications tools of cell phone and satellite easily vault over mountains and rivers. Our information technologies pose a uniquely twenty-first-century danger to the integrity of the wild as the latest leaps in technology threaten to shrink the *mental space* provided by wilderness.

Exhibit A: Google is busy making plans for what has been called "universal connectivity." The information technology giant is expected to spend between $1 billion and $3 billion to deploy a fleet of 180 mini-satellites that will provide an Internet signal from the sky. The satellite connection may be augmented by high-altitude balloons and/or solar-powered drones supplying high-speed, broadband service. Someday soon, all of Earth might be a wi-fi hotspot. You'll be able to check your e-mail and update your status from the farthest reaches of the bush.

Google is also engaged in an ambitious effort to photo-map some of the world's most remote places, including wilderness areas. In the spring of 2014, the company unveiled "Google Treks," an extension of its popular Street View program. As part of its "quest to map the Earth," Google has sent explorers to Australia's Great Barrier Reef, the Galapagos Islands, and Volcanoes National Park in Hawaii equipped with backpack-mounted, 360-degree, fifteen-lens arrays to photograph remote scenes. Thanks to Google, you can now raft through the bottom of the Grand Canyon and travel the "road to nowhere" in the Canadian Arctic without leaving the comfort of your laptop.

Perhaps these are modest, well-meaning domestications. Universal connectivity could provide Internet access to billions of people who have never experienced its promise. Photo-mapping the wilderness might provide biologists with important data, or give disabled people the chance to see places they otherwise never could.

But make no mistake: those technological aspirations are domestications nonetheless. It's a kind of taming by ones and zeroes that threatens to make wilderness obsolete.

Every generation has its own peculiar anxieties about technology. The twentieth-century wilderness movement of Aldo Leopold and Howard Zahniser was a reaction to the rise of the automobile. And I'll bet that when the first telephone was installed in Yosemite Valley, someone called it a sacrilege. Each generation's complaints about technology eventually seem quaint to its successors, and I'm sure that when every backpacker is wearing some kind of computer embedded on wrist or forehead, my rant here will be a charming anachronism.

Still, the impacts of today's inventions are different from those of past generations, if only because the velocity of inventiveness has increased. Thanks to Moore's Law (which says that computational power doubles roughly every two years), future shock is now a chronic condition; the technological baseline shifts every time Apple comes out with a new gadget. To understand how quickly technological change is occurring, consider this: we have just barely started to wrap our mind around the Age of Man, and already some Silicon Valley seers are trumpeting the impending arrival of the "post-human" era. Some futurists predict that we will soon meld the hardware of the human body with digital software to create augmented (or "improved") humans and, in the process, make a leap forward to a species beyond *Homo sapiens*. Other futurists imagine something even more grandiose: supposedly, by the middle of this century we will arrive at what has been called "The Singularity" as we begin to upload individual consciousness into computers to achieve a godlike omniscience and immortality.

These techno-utopian fantasies should not be dismissed or underestimated. Some of the smartest minds in America are hard at work making them into reality. To me they reek of insanity. If we ever do achieve the everlasting life of synthetic intelligence, it will mark the final, perhaps irrevocable, departure from our birthplace in nature. At that point, it won't even matter whether Earth is still habitable for humans—we'll no longer live here. We'll be residing somewhere in the mainframe.

If there's any antidote to the fever dreams of those working for a cyborg future, it seems to me that it resides in the plain old embodiment offered by close contact with the wild. I think you know what I mean, the *feeling* of wind on skin, of icy ground underfoot, of a hard rain coming down. The visceral experience of wild nature—its implacable physicality—acts like a splash of cold water to the face, bringing us back to our five senses.

In the wilderness, forced to grapple with the uncompromising elements, we are sometimes reminded of original meanings. A *web*, for example, is something you get stuck in. A *net* is designed to catch and capture.

Once again around the fire, sharing stories in the dark.

Matt was telling us about an experience he had while "on Project." He's the kind of guy who speaks carefully, each thought measured. As he talks he sometimes waves his hands back and forth, as if he were swimming through his words as they leave his mouth.

"One day, I was walking, just wearing a loincloth. On instinct I started to follow the curve of land. And I came to this beautiful waterfall, and I climbed up around it, and came to this small, grassy meadow that you could only reach from one direction. It was perched above this lake. At the edge of the meadow there was a field of boulders, rising into a massive wall of stone.

"A small stream ran through the meadow, curving back and forth. I knelt and put my face in the stream. And it was like I touched something ancient. There are no words for it."

The wild does, indeed, resist definition. I've walked hundreds of miles over mountains and through canyons and across forests and deserts trying get closer to whatever that ineffable essence of wildness is. And I've kept on walking because—I guess like Lynx, Matt, and Jessie—I believe that in wildness resides a vitality missing from our domesticated lives.

A gardener quickly learns about wild vigor. My years at Alemany Farm—three and a half hand-worked acres and never enough bodies—showed how insistently the wild edges back. If I neglected a plot for a season, the next thing I knew I'd have a patch full of weeds. Clinging purslane or deep-rooted malva (both edible "weeds") easily overwhelm a row of *rouge d'hiver* lettuce. Without yearly tending, the footpaths between the vegetable beds get tangled with morning glory.

"Nature has a place for the wild Clematis as well as for the Cabbage," Thoreau wrote in his "Walking" essay. She does. But if you're not careful about keeping some space between the two in your dooryard garden (probably via a hell of a lot of pruning), that clematis will quickly vine around your cabbages and choke them to death.

There's simply something tougher about wild things. A wild mustang is craftier than a horse that has been—I've always hated this expression—"broken." A coyote is wilier than your average dog. I've kept chickens, and I've been grateful for the years of eggs and (eventually) weeks of soup they fed me. And I know that in intelligence and wit a barnyard hen is no match for a scrub jay.

Somewhere in the kernel of our consciousness we've always known that the wild has a unique power. How else to explain the many cautionary tales—Adam and Eve's fall from grace and Prometheus's heinous crime and the sadness embedded in the Epic of Gilgamesh—involving some kind of epic sundering of humans from wild nature? Even now, in the Human Age, our departure from the nomadic, hunter-gatherer way to the sedentary lifestyle of the farmer remains the biggest rupture in the human experience. The leap from the Paleolithic to the Neolithic changed our relationship to the rest of life like nothing else. In comparison, the much-celebrated move from the agrarian to the industrial (still under way in much of the world) just replaced the intimate drudgery of the farmyard with the alienated drudgery of the factory. And the transition from the industrial to the information age remains a work in progress; we still haven't figured out how to substitute electronic bits for actual things. Few other inventions separated our species from wildness like the blade of the plow. Even a committed agrarian like Wes Jackson—founder of the Land Institute in Salinas, Kansas—speaks of agriculture as the first sin, the act that ended Eden.

Recent plant science confirms the insights of ancient myths. Researchers have found that as humans domesticated fruit and vegetables plants, we have bred much of the nutrition out of them. Over millennia, farmers selected for plant varieties that were less bitter and sweeter, while at the same time lower in fiber but higher in starch. That is, we bred for energy density, but at the expense of nutrients. The result? Spinach has seven times less phytonutrients than wild dandelion, phytonutrients being the compounds that health researchers believe are linked to lowering the risk of cancer, heart disease, diabetes, and dementia. Heirloom purple potatoes from Peru have twenty-eight times more beneficial anthocyanins than your typical russet potato. "Unwittingly, we have been stripping phytonutrients from our diet since we stopped foraging for wild plants some 10,000 years ago and became farmers," according to Jo Robinson, author of the book *Eating on the Wild Side: The Missing Link to Optimum Health.*

I am unaware of similar science on wild meats versus farm-raised meat. But all the wildlings were convinced that there was something more—well, *vital*—about eating game. "Energetically, I think, there's something that happens when you put wild foods in your body," Lynx said. "I know that everything has an energy. If we feed ourselves with the energy of the wild, we're putting the wild inside us."

Matt told me, "There is something there in the wildness of that plant life or the life of that wild animal that makes me appreciate it so much more—if I am given the gift of its life for my sustenance. Maybe it's just that appreciation. Maybe it's the placebo effect. Or maybe it's something real. To me it feels different to be eating that food. And it tastes very different. And it feels much more . . ." He struggled for words, then said, "More *enriching*. It feels more potent. More powerful, I guess."

I admired the wildlings' courage in pursuit of the primal. I, too, wanted to eat food that was more "potent." I envied how feral they had become, how virile. But I was pretty sure I would never be able to follow them. Domestication is a habit of mind, and I'll admit that although my aspirations might be bohemian, my impulses have always been bourgeois. Even in the middle of the wilderness, I like my creature comforts. I like my coffee and my lightweight, folding camp chair. I like the space fabrics that keep me warm and dry. I appreciate the modern magic of a lighter.

Still, when I think of all of the crap that I haul into the backcountry, it makes me wonder: *Who's really playacting—Lynx or me?* Because I know that all of the gear that gets me to paradise is dependent on depredations someplace else. The gas for the propane cook stove likely depends on fracking. Without a petrochemical plant somewhere, there's no plastic to wrap the food in.

The Stone Age Living Project succeeds by showing how, through using only primitive tools in the wild, you could ease that cognitive dissonance. You could patch the age-old break between wild nature and human technology. But only for a short time. The experiment is just that—an experiment.

The open secret at Lynx's place was that the 2014 Project had been, in some sense, a disappointment. Or, at the very least, a sobering wake-up call. The experience had revealed how impossible it is to "Live Wild" for any extended period of time. Yes, Stone Age living could soothe the tensions between wild nature and human culture, but only as a getaway.

Matt is a Brown University–trained engineer, and he referred to the Project as a "proof of concept," one that illustrated the challenges of attempting to return to the Pleistocene. During their twenty-six days in the wilderness, the wildlings had never gotten close enough to a deer to take a shot with their handmade bows. They caught only a handful of fish using bone hooks. With the exception of foraged greens, all of their food had been gathered beforehand and carried with them. While the diet of dried buffalo meat, dried mushrooms, dried berries, and liberal dollops of preserved deer fat had felt filling, it was somehow less than fully nutritious. Many people complained that their stool was heavy and hard. Matt said he had never felt so weak and foggy-headed in his life.

"We could go out there and do it for a month—but we could not harvest enough food to sustain ourselves," he told me. "That was a crux point. It was realizing we couldn't sustain ourselves indefinitely. There was a clock ticking. At one point, Jessie said, 'Is this vacation or is this life? If it's vacation, it's okay. But if this is life, we're starving to death right now and we'll all be dead in two to three weeks.'"

Jessie later elaborated the point for me. "'Living Wild' is tricky, because out there we were dying, we were starving as a clan. And that was sad, because we weren't living. We were visiting—and that was disappointing."

None of this is news to Lynx. Having led the Project for ten years, she understands the contradictions of Stone Age living better than anyone. Lynx still has the self-deprecating vibe of the punk she once was, and she's possessed of a healthy sense of irony ("a primitive girl's handbag" is how she mockingly referred to one of her fur pouches). She is under no illusions about the limits of the Project, which is sustained in no small part by roadkill and hand-me-down hides donated by area hunters.

Lynx told me, "People hear about what we're doing, or they see the pictures, and it looks very romantic—and it is. I'm absolutely a Romantic. But it's also very real and very harsh and very unforgiving. And very challenging. And it's not something that you can just jump into. The earth doesn't give a shit if you die out there. Or, you'll do a lot of damage out there. It has to be done conscientiously. Otherwise, we'll have everyone going out and hunting and fishing and we'll wreck the planet. We can't all do this, because there's too many of us."

As everyone understood, the core problem of the Project was the need to be remote and apart from civilization. The only place where you can attempt to pull off Paleo is a place that's also unfit for long-term human habitation—the alpine landscape of the high country. Alpine ecosystems are often areas that biologists call "depauperate." That is, there's not much wildlife there—or at least not much *large* wildlife, which is to say, game. Lynx summed up the predicament in one of her journal entries from the Project: "This high country is not a place for people to live. To journey through, to hunt, to pray on the bare, windy mountaintops close to the gods—yes. But not to live. The river valleys are the places for the people to live, where the roots and the berries and the fish and the deer are more abundant." Her journal entry for that day closed with a lament for civilization: "What have we done? Forgive us. What have we done?"

What we've done is taken the best places for human homes. The wilderness conservation movement has been criticized for mostly protecting "rock and ice." At first this alpine fetish was a Romantic hangover, but it soon became a political necessity: rugged, remote peaks were some of the only places remaining that hadn't succumbed to human development. We built our cities in the areas endowed with the greatest natural wealth—New York City at the mouth of the great Hudson River, Seattle along the salmon-rich deltas of the Salish Sea, industrial centers ringing the shorelines of the Great Lakes. At the same time, we put our farms on top of the once-great plains. America's tallgrass prairie is all but extinct, smothered under a carpet of corn. Across the continent, the most abundant places have all been paved or plowed.

This is the unresolvable dilemma of the Stone Age Living Project: the only spots where one can have the space to live wild are the places where there's no food. And so the age-old script repeats itself—the nomads who are committed to hunting and gathering are pushed once again to the margins by civilization.

Lynx is working to address this challenge by securing a larger piece of land. She's in talks with private landowners in Montana and New Mexico about using their property to go Stone Age for a longer period of time. She figures she needs at least 30,000 acres of wildlands. Someplace with sizeable populations of either elk or bison would be ideal. If she can secure such a site, she would like to attempt a yearlong Project with a clan of already-tested wildlings. She imagines that it would be the world's first "Primitive Human Preserve." If Lynx pulls it off, it would make for a tidy twist of history, finally fulfilling painter-conservationist George Catlin's original North American preservation ideal of a land reserved both for animals and for people.

The Project would still be an experiment, an island of the Pleistocene in a sea of the postmodern Human Age. But it would be a major accomplishment nonetheless—a powerful bit of performance art that could tweak prevailing views about civilization's relationship to wild nature. A Primitive Human Preserve would be a living reminder that "progress" does not move in one direction, and never has. There are no straight lines in nature, and progress is no different. Sometimes what we think of as technological progress also diminishes human well-being (think: chemical wonders vs. cancer). And often as not, social progress (that is, the expansion of human liberty) is random, shaped by stochastic disturbance, as ecologists would say. Many paths are available to us—trails that lead not necessarily backward, but sideways or at some unexpected angle. A Primitive Human Preserve would be a wilderness that's not just a biological refuge, but also a last resort of the imagination. Keeping the wild is a way of leaving our options open to the unexpected twists of the future.

"Wilderness is a state of mind," environmental historian Roderick Nash has written. The observation illustrates the importance of preserving the wilderness for civic reasons as well as ecological ones.

At this point, to know that some lands remain outside the matrix—that we still have a true Away to which we can escape—is a necessity for political liberty.

Go back to the OED one last time. *Wild*: "not subject to restraint or regulation." Also, "rebellious."

As Gila-area hunting outfitter Brandon Gaudelli had reminded me, to be wild is to be free. Even in a mostly domesticated society, a dose of wildness—whether psychological or physical—is an essential condition for political freedom. Wilderness's autonomy, its sheer self-will, frustrates the dictator and the king. So it has always been. The ancient Greeks' wild Dionysian festivals were a time to flout convention. In Norman England, "the woods" were a place outside of the law. As the eminent ecological historian David Worster has observed: "Nature in its wilder state is a threat to authoritarian minds."

The idea of the wild as an important ingredient of equality and freedom grew naturally in North America. Emerson, for one, understood the connection between wild nature and an open society. "In a good lord, there must first be a good animal," he wrote in his *Second Essays*, "at least to the extent of yielding the incomparable advantage of animal spirits." There's an ecological awareness embedded in the line, a recognition that a healthy ecosystem is a republic without tyrants. There's a suggestion of bio-mimicry, too, an encouragement for us to look to the elegant chaos of the untamed to order human affairs.

As his model, Emerson had not just the North Woods of New England, but also the vanishing example of Indigenous societies. When Europeans arrived in the imagined wilderness of the New World, one of the things that most amazed them was the egalitarianism of the societies they encountered. "Every man is free," an explorer told an astonished British audience, discussing Indian life, and no person "has any right to deprive anyone of their freedom." Even among the politically sophisticated Iroquois—famed for their multi-tribe constitution—there was no such thing as "a chief." There were, to be sure, headmen and matriarchs, people who wielded great influence and gathered followers. There were taboos. But there was no compulsive authority. Decisions were made by the consensus of the council meeting, and no man could force another man to do something against his will. Such small-a anarchism—or, if you prefer, communal libertarianism—was a signature of pre-Columbian societies across North America, from the Cherokee to the Lakota to the Apache.

Many scholars argue that this model of political equality was among the most important trades of the Columbian Exchange. A living example of free and democratic human relations proved just as contagious as smallpox. Moving in the other direction, it infected the colonists and their cousins in Europe with new ideas about the political good. The wildness sparked at the Bastille in 1789 came, in part, by way of America.

In America the persistence of places beyond the ax and the plow has contributed to a culture of liberty. Freedom requires not just openness, but also capaciousness—a sense of the world as large and wide. The wild is a guarantor of liberty because it serves as an actual, physical escape. This ability to get away has been an essential part of the American experience. In US history the wilderness has been a last resort for the apostate, the nonconformist, and the fugitive slave. Empty, wild country gave Crazy Horse a place to go off alone and find the vision to lead his people.

The mid-twentieth-century minds behind the American conservation movement instinctively understood the relationship between wildness and freedom, having just witnessed the totalitarian horrors of the Second World War. In his "wilderness letter," Wallace Stegner wrote: "If the abstract dream of human liberty and human dignity became, in America, something more than an abstract dream, mark it down at least partially to the fact that we were in subtle ways subdued by what we conquered." Which is a poetic way of saying that the wilderness changed us. The open vistas of wild spaces helped mold the American spirit of fierce independence by providing us with what Stegner called "the geography of hope."

That anarcho-wildman Edward Abbey made the point forcefully, writing: "We cannot have freedom without wilderness, we cannot have freedom without leagues of open space beyond the cities, where boys and girls, men and women, can live at least part of their lives under no control but their own desires and abilities, free from any and all direct administration by their fellow men."

Here now in Anthropocene—all-encompassing and all-consuming and, therefore, with a totalitarian vibe about it—the wild's civic value may be its most important virtue. As a political necessity, we need to keep some places outside the reach of our awesome new technologies.

Big Data in the backcountry? No thanks. Wi-fi in the Woods? I think I'll pass. Because if we ever succeed in knitting all (or even most) of the physical world into the Internet, we could end up abolishing the sense of the Away. And we need the Away, as a political good as well as a psychological one.

Our Human Age is characterized not just by human omnipotence, but also by our civilization's attempts at omniscience and omnipresence. Certain cultural values are embedded in every technology: the assembly line is about efficiency and uniformity, the automobile expresses the desire to conquer distance. If there is any cultural value inherent in the Internet, it's the wish to see and know everything. We are almost everywhere, what with our constant connectivity and our Google Trek. But our amazing information and communication technologies threaten a kind of digital enclosure that is every bit as inimical to freedom as the enclosure movement centuries ago in Europe that pushed peasants off the land. The litany of incursions into our private lives is all too familiar: thousands of cameras mounted throughout our cities, real-time tracking of our cell phones' locations, government sweeps of e-mail, corporate monitoring of every web search and credit card purchase. It's not an exaggeration to say that, in our digital lives at least, we're living in the Panopticon.

I'll admit that, in some ways, omniscience is pretty awesome; I like as much as anyone being able to Google, say, the population of Reykjavik. Omniscience, however, doesn't jibe with the essence of wilderness—a place still governed by mystery and wonder. The wild is unknowable, if only because most of the action there passes unseen. Or, as the Lakota would say, the wild is *wakan*—incomprehensible.

We need to defend the wild, then, because in an otherwise programmed and micromanaged society, it remains one of the last bastions of unpredictability. For now, the wild remains an oasis of anonymity in a world in which we are constantly tagged, pegged, and followed with digital breadcrumbs. In this era of NSA's PRISM and the constant tracking of Big Data, having a few places that are disconnected and unmonitored seems more valuable than ever. And so, among its many other tasks, the twenty-first-century conservation movement will have to commit to maintaining the wilderness offline, as a place where citizens can walk unwatched.

The wildlings at Lynx's camp understand this better than anyone. When I asked why she had spent a year at Living Wild learning ancestral skills, a woman named Epona told me, "I wanted freedom. I wanted to stay sovereign. And the only way I could do that was to go to the wildlands, the last public domain." She was wearing long black earrings made out of buffalo hair and cooking her supper over an open fire. A raccoon pelt was hanging in a nearby tree. She said, "If I can make my own gear, I am more sovereign." *Sovereign*—that is, the ruler of herself.

Thoreau understood all of this back when the telegraph and the railroad were novelties. "Walking" begins: "I wish to speak a word for Nature, *for absolute freedom and wildness*," as I can't help but emphasize. The essay, remember, is a celebration of the saunter—to stroll without destination or direction. To have the space to saunter is to be free.

When I think about the wilderness as a civic good, Thoreau's famous dictum—"in wildness is the preservation of the world"—takes on yet another layer of meaning. Perhaps it was not written by Thoreau the naturalist or Thoreau the poet. Perhaps instead it was written by Thoreau the tax-resister, the political philosopher, the dissident.

I finally got a fire started by rubbing two sticks together. It was a wet, cold morning, and we were huddled in the Fire Lodge, eager for warmth. Each of us was working a bow drill or hand drill and trying to birth a coal when mine caught first. It felt great watching the flick of the flames, like I had cracked some ancient code.

We spent the rest of the day making knives or needles and sheaths or purses out of the bone and skin of deer forelegs. This was to be our final lesson of the course. In the days since the fire-starting instruction, we had learned the basics of primitive shelter building, glue making from pine pitch and deer hide, as well as how to set a deadfall. (We caught one mouse and one chipmunk, which together made about half a meal.) A whole day was given over to instruction about foraging medicinal and edible plants in the Eastern Cascade ecosystem. On the last night together we had a celebratory wildfoods supper of venison steaks, a mallow and dandelion and nasturtium salad, hawthorn berry cakes, and a rosehip and wild apple tea.

I loved every minute of it. But I couldn't get away from the thought that Paleo living is an impossible model for humanity today. Epona's partner, Alex, pretty well summed up the conventional wisdom at the Living Wild School when he said, "Seven billion people can't all live this way. Probably the population of the United States can't live this way."

No, they can't. Which leads perhaps to a uncomfortable conclusion: to protect what remains of the wild, we will have to commit to taming ourselves.

If we truly want to keep some places autonomous and self-willed, we'll have to domesticate ourselves further. To share space with wild plants and animals will require that we shrink humanity's footprint. This means, among other things, that we'll need to have fewer babies and finally find a way to slow and then reverse human population growth. We'll have to staunch our ceaseless consumption of the planet's resources, returning to the old ethic that valued craftsmanship over quantity. More of us will have to live in cities, and those cities will have to become denser and taller. We'll have to ditch the convenience of our cars for the camaraderie of the train and the bus. At the same time, we'll have to further intensify our agriculture and grow more food on less land. We'll have to use every electron more wisely. Those of us lucky enough to live in the wealthy nations will have to do much better at sharing the planet's riches with the billions who remain energy-poor and calorie-hungry. In short, we'll need to rein in our appetites and restrain our baser instincts.

For those of us raised on the romanticism of going "back to the land," it's a tough contradiction—recognizing that the clearest path to preserving the wild is by further taming the human spirit. But there it is: only by shrinking back can we allow wild nature to take back some more of the land it needs to thrive. Accepting this truth is part of the hard labor of forming new habits of thinking about wilderness and human civilization. It's difficult in the way that hope usually feels difficult, when our heads tell us that the situation is all but hopeless.

Still, I'm glad that Lynx and the wildlings are out there. It makes me happy to know that some people still follow the old ways as best they can, moving in circles like the nomads of yore, hunting and fishing and foraging. I like seeing that there are still some people living close to the earth—living not "off" the land, but *with* it. The knowledge is a consolation of sorts. It gives me confidence that, even as the gloaming appears to deepen, someone is still carrying the torch.

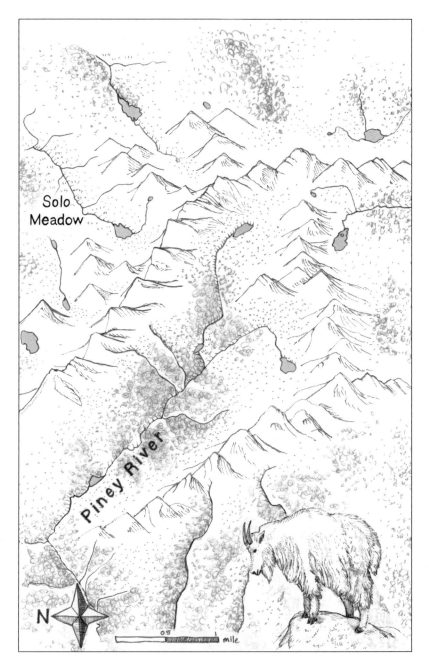

Solo
Meadow

Piney River

N

0.5   1 mile

Eagle's Nest Wilderness

Epilogue

Wild at Heart

W<small>E SCRAMBLED OUT OF OUR SHELTERS</small> before the sun topped the eastern peaks, our breath visible in the crisp dawn air. The full moon was still in the sky. It hung above the sharp point of a no-name mountain, the bare rock turning peach now in the first light. Snow patches clung to the north face of the ridge. A small meltwater pond mirrored the scene, doubling the spareness above treeline.

The boys began making their breakfast. Day 18 (or was it Day 19?) of porridge or cold bagels or crappy muesli. While some of them prepared the food, the others got busy breaking down their simple tarp tents, all the while nudging each other with the kind of casual insults that, among adolescent males, often passes for affection.

Back to the High Country. This time to the Rocky Mountains, the kind of alpine scenery I'd been taught to appreciate, the first modern wilderness. I had gone into the heights of Colorado's Gore

Range to explore how the wild might affect a group of young people unencumbered by theory. I had covered too much ground, had gone into too many intellectual box canyons and side streams looking for a twenty-first-century wilderness . . . and perhaps I had gotten myself lost.

Maybe what I needed was to hear it all fresh, from the next generation open to finding something worth saving in primitive places. After all, today's kids are the ones who will have to make a home in the Anthropocene. They'll have to figure out where—or whether—wilderness fits in civilization's twenty-first-century survival kit.

The day before, the eight boys—ages seventeen to twenty-one—had completed a three-day "solo" as part of their twenty-two-day Outward Bound mountaineering course. Each of them had been alone for sixty-seven hours, with nothing but a tarp for a shelter, a sleeping bag, water, and a scant amount of food. Nine Bickies crackers (think, Saltines without the salt), a packet of raisins, a packet of peanuts, and some electrolyte mix. A total of 1,000 calories for three days. It was archetypal trial-by-solitude—to search for a new self-awareness alone in the wild. Muir would've been proud. "Only by going alone, in silence, without baggage, can one truly get into the heart of the wilderness," he once wrote.

I had been waiting to meet the students when they came off their solo. A day earlier I had packed into the mountains with a couple of longtime Outward Bound instructors—Nate and Dustin—to rendezvous with the group. The trip to the group's site had been tough for me. I may be an experienced trekker with a decent amount of woodcraft, but mountaineering is not my thing. When Nate and Dustin pointed to a line of mountains and said we were going *over* the heights instead of *around* them, I hoped they were joking. Then we made the climb: 1,300 vertical feet slogging up and over a wall of rock, then down an ice chute where we had to use axes to make the descent. I later joked with the guys that I wasn't half as scared as I had looked coming down the ice chute—I was *twice* as scared.

But I was glad for the ass-kicking. The experience had given me a sense of what the kids had been through, how tired they must have felt after two-plus weeks in the backcountry. My sore legs gave me some compassion for their bewildered expressions when they stumbled in from their solos and plopped down on their backpacks, desperate for food.

After some lunch, we had made a short climb to the lake where we were now breaking camp. There, atop what seemed like the roof of the world, the boys sat in a rough circle talking about what it had been like, being alone for that long in the wilderness.

Jon, a hockey jock from Philly, said he was surprised by how difficult the experience had been. During his solo, he said, "I realized how frail I was." Nathan, a computer geek with a mop of ginger hair falling around his glasses, said he realized that so many of the things he takes for granted at home ("lights, heat, plumbing") are luxuries. "It was enlightening," he said in a soft voice. "It gave me a fresh appreciation for life. For existence, I suppose." Brandan, a dude from Plano, Texas, who had been sent on the trip by his oil-executive stepfather, had felt something similar. "It's like, on this one planet, there are two separate worlds. It's just totally different. Being out here is like being on another planet."

I knew exactly what Brandan meant. In comparison to our cities and farms, the wilderness does feel fantastical: the morning moon like a scene from another solar system, the calm of the high mountain pools like something from a dream. Yet the observation also made me sad. What a shame, to think that our own Earth has come to seem otherworldly. Once commonplace sights and sounds—the stars at night, the burble of a stream—are now curiosities.

A discouraging thought, but not exactly a novel one. There's a lot of talk these days about how young people are alienated from wild nature. A journalist-turned-activist, Richard Louv, coined a phrase for the phenomenon—"nature deficit disorder"—in his book *The Last Child in the Woods*. At the same time, many environmentalists are (rightly) worried that the ranks of wilderness lovers is not only too old, but too white. A long history of exclusion makes it difficult for some people of color to see parks and wilderness areas as inviting spaces, and so they sometimes avoid them. As Rue Mapp said to me our last morning on the tundra, "We need to create dotted lines between the wilderness and the places people live, whether that's Oakland or Brooklyn. Because a lot of people see them as disparate."

Statistics illustrate the challenge. Seventy percent of those who regularly engage in outdoor recreation are white, a disproportionately high number. This is worrisome for obvious reasons. As the United States becomes ever more ethnically diverse, there may be an ever smaller constituency of wilderness aficionados, which could eventually jeopardize the political commitment to public-lands preservation. In one of our conversations, Lynx summed up the problem: "You can't love what you don't know, and you won't fight to protect what you don't know." *The wilderness?* For multi-ethnic millenials, that's just a bunch of old, dead white guys tromping through landscapes that no longer exist.

"I find that almost no one I know who is forty or younger goes backpacking," a Moab-based journalist, Christopher Ketcham, has written. "This is a kind of heartbreak." Yes, it is. I also worry that backcountry recreation is becoming an esoteric art—the twenty-first-century version of those medieval monks off in the hinterlands, illuminating manuscripts. A craft, known only to a few, that seeks to preserve a body of knowledge and truth via beauty.

Those, at any rate, were some of my rather gloomy thoughts that morning as I hoisted my pack and got ready to make the climb out of the alpine basin, the first ascent of the day's two mountain passes. To cheer myself up, I remembered something one of the students had said the evening before. It was André, a jolly giant of a kid who, just a couple of years earlier, had moved with his two younger siblings from Haiti to Dorchester, outside of Boston. Although he was often laughing and cutting up, André's English was a work in progress, and he didn't speak much. In the circle he had been the last to talk about his solo experience. In his thick Creole accent he said, "It gave me an opportunity to explore how the nature is good for me. I was surprised that I wanted to explore the nature more. Just being outside."

The Gore Range of the southern Rockies is a landscape of high drama. As the crow flies, the boundary of the Eagle's Nest Wilderness is only a half-dozen miles north of the fancy resort of Vail.

But because the area is so rugged, and because much of it is trail-less, few people go there. From the lower slopes of lodgepole and aspen, the mountains rise sharp, swift, and seemingly impenetrable. Imagine pyramids stacked on top of pyramids, the stone arranged in rough-edged triangles until they finish as peaks. On the mountains' shoulders are groves of blue-tinged spruce, aquamarine lakes, and long, flower-flecked meadows. By August the scene is an explosion of color. Millions—tens of millions!—of mauve daisies, magenta paintbrush, and yellow alpine sunflowers. Columbines, mountain bluebells, star gentian, alpine clover more intricate than the fanciest hybrid tulip. Streams spilling down the mountainsides like braids of silk.

Beautiful though it is, the Gore can be treacherous. Ledges drop off into sheer cliffs where the snow holds out in the shadows all year round. Easy footing can be hard to come by. The peaks shed huge talus slopes that a mountaineer has to cross carefully: boulder to boulder, slab to slab, knowing that a false step could lead to a broken bone—and that bone could be your neck. The summer afternoons are storm-prone, either a violent burst that lashes the mountains quickly, or else an hours-long drizzle that soaks you to the skin. If you're making a peak attempt or negotiating a high pass, the dark clouds are trouble. You don't want to be out in the open on a mountaintop when you hear that rumble of thunder followed by a lightning flash.

In such a raw territory people either come together in a common task—the trail as bonding agent—or else they fall apart. The group of boys I was following? They were falling apart.

I had landed myself amid a band of lovable knuckleheads, like *The Bad News Bears* of the backcountry. This wasn't a crack team of alpine commandos—just a motley crew of young men on the precipice between childhood and adulthood. They spent most of their time debating the finer points of fast-food cuisine: Church's Chicken versus Burger King, Chipotle versus Buffalo Wild Wings, how much pizza they would eat when they got home. Maneuvering across a talus slope or huffing up a mountainside, I heard many awful attempts at freestyle rapping. This being an all-boys group, there were also fart jokes—a *lot* of fart jokes.

I was in high school when I first heard about Outward Bound, and though I was intrigued, I figured it was just for rich kids, something out of my parents' financial league. Of course, wilderness recreation has always had a reputation as a wealthy person's pastime (remember, it was Manhattan Brahmins like Roosevelt and Alaska aficionado Bob Marshall who pioneered the American wilderness aesthetic). So Outward Bound has fought hard to overturn the stereotype and make sure to recruit low-income students. The group I was following was pretty representative for a Colorado Outward Bound course. Many of the students had grown up with real wealth, the sort of kids who have friends with movie theaters in the house. But a full 40 percent of the students were on scholarship. Everyone agreed that the group had an impressive degree of cultural diversity.

Adeyemi (or Ade, like *a-Day*,) had come from London. He was born in Lagos, Nigeria, and moved to the United Kingdom with his mother and sister when he was eleven. They lived in council housing in Lewisham, East London, where Ade distinguished himself in his mostly immigrant school. With his Harry Potter–like glasses and steady calm even in a landscape that he admitted "petrified" him, Ade was an obvious overachiever. He had turned down Oxford, he said in his clipped English accent, because he thought attending university in the States would be more challenging.

Pat grew up in Bethesda, Maryland, his parents in real estate financing, and he had attended Sidwell Friends, the same DC private school that the Obama daughters go to. He was physically fit (a cross-country runner) and outdoorsy (he had gone backpacking with his parents) and had proven himself to be a steady member of the group. As had Nathan, a nerd's nerd who was into building video games and hanging out with his medieval sword-fighting club. Nathan's steadiness made him something of the opposite of Brandan—a Texas jock (safety on the football team, shortstop in baseball) who had struggled with ADHD and who seemed a bit adrift.

Either Jon or Seni could have been the natural leader of the group, but it hadn't happened. At twenty-one, Seni (a nickname for Jens) had the most life experience. He had served two years in the Norwegian military, and now was headed to Notre Dame, where he planned to major in finance and then, he hoped, to work in the City of London. He was physically and mentally strong (by far the best at map-reading and navigation), but he admitted to me that by Day 14 he had become so tired of the bickering in the group that he had backed off, content to carry his weight and nothing more.

The natural charismatic in the group was Jon, who easily could have been cast as the cool kid in a John Hughes movie, or, for that matter, a Justin Bieber stunt double. When I saw his prep school T-shirt for an all-boys school outside of Philadelphia, I pegged him as a rich kid. Then I learned that he was the only one who had paid the $4,100 course fee out of his own pocket—after working for five months at a cement quarry. His charisma would have been an asset were it not for the fact that his mercurial moods kept the whole group on an emotional roller coaster.

Someone always has to be the slow-poke, and that was André, the Haitian kid. By the time I arrived, he had already lost about fifteen pounds, but he was still carrying what looked to be some extra weight, and the group often found itself slowing down to accommodate his shuffling gait. This made André feel bad about himself, and his normally sunny attitude would be clouded by a dark mood as he marched along sullenly.

And then there was Will, who was sort of like an extra chili in an already peppery stew. He was a big seventeen-year-old from the expensive suburbs of Northern Virginia, his father a former White House official. But Will was a headstrong kid who had dealt with his own personal challenges. He had ADHD and Tourette syndrome, which at one point had been so bad that his involuntary spasms could crack a school desk. The time in the wilderness had clearly been good for him. His Tourette's, he said, wasn't as bad in the backcountry, which he had found to be "serene" and "tranquil."

Oh, and I have to mention Winston and Sheila. "Winston" was the boys' name for a powerfully built mountain goat that had been following the group for more than a week. The goat had shadowed them over one mountain and another, eager for the salt in the group's urine. Eventually the billy was joined by a nanny that the crew dubbed "Sheila." The pair of goats was a constant presence, sometimes coming to within yards of us.

Individually, they were all good guys, but together the group was a mess. The course was just days from finishing, and yet there was little cohesion or trust among the boys. One would go route-scouting a cliff edge, and within a minute another kid would be questioning why he was up there and when he was coming down, and then a third would demand to know where they were going, and a fourth would ask how much farther, and soon the whole crew would be embroiled in an argument.

I was following the group through what was supposed to be their "finals"—three days in which they would guide themselves through the mountains while the instructors stepped back. But the students were working together so poorly that the instructors hadn't been able to fall away and let them be on their own.

The first day of finals involved a grueling, twelve-hour march over countless talus fields, an instructor-assisted push over a spot called "Kneebuster Pass," and then a late camp set up amid a threatening storm. The next day of finals was supposed to be a more or less straightforward trek through a valley bottom split by Piney River. But the students were lost by mid-morning. When they arrived at a crossroad in the trail, there was a debate about which direction to go. Nathan closed off the discussion by tapping his temple and announcing confidently, "I've got it up here." At which point they went exactly the wrong way. (For the record: I knew as much, but kept my mouth shut.)

Disoriented, they then had made a series of bad decisions. The group unnecessarily went off-trail, climbed down a steep granite slope, and made a dangerous and thoughtless river crossing. Then they started bushwhacking in a direction about 120 compass degrees from where they should have been pointed.

The instructors had to jump in. The three trainers hurried down the slope, yelled for the kids to stop, and then had to wade them back across the river in a wet-foot crossing. The students were frustrated. The instructors were disappointed and exasperated. It was time for a talk.

The lead instructor, Vince, pushed his hand back through his hair, sighed heavily, and said, "You guys aren't working as a team.

You're not listening to each other. You don't know where you are. You're taking unnecessary risks. To make it out of here, you need to work together."

The students mumbled among themselves, no one offering much of a defense. Then Nate, one of the veteran Outward Bound instructors who had guided me into the mountains, stood up in front of the kids and delivered this speech:

> Outward Bound started as a survival skill because young men— boys sixteen, seventeen, eighteen—were dying in life rafts at the hands of the Nazis. And why were they dying? They were dying because they weren't working together. Because of ego. They weren't taking care of themselves. They weren't taking care of each other.
>
> If this is the hardest thing you've ever done in your life, I have news for you: you're really fucking lucky. Because life is about to get a lot more real for y'all.

The wilderness of Outward Bound is not John Muir's woodsy getaway or Aldo Leopold's ecological preserve. It's more of a Teddy Roosevelt, Alfred Lord Tennyson wilderness, a terrain in which to undertake a savage test of self. In the Outward Bound way of thinking, the wild is a crucible of character.

As Nate said in his riverside exhortation, Outward Bound began during World War II, when the US Navy realized that young sailors whose ships had been torpedoed in the dangerous North Atlantic crossing were dying in large numbers even when they managed to make it into a lifeboat. They were dying because they lacked basic survival skills, and also because they were unprepared for working together in an environment—in that case, the open ocean—stripped down to its basic elements. Ironically, the school's founder was a German, Kurt Hahn, who in the interwar period had founded an adventure school in the Scottish Highlands designed for the British upper class and dedicated to erasing what he called "the enervating effects of privilege." After the war, the program was expanded to civilians; the Colorado Outward Bound School led its first wilderness expedition in 1961.

The school's insignia is a nautical flag called the Blue Peter. It's a gray square on a field of blue that boats may hoist when headed from the harbor to sea, literally "outward bound." It's a symbol for a journey into unknown perils. Hahn was clear, however, that his course was not adventure for the sake of adventure—nor merely survival school—but instead was about "value-forming experiences." Hahn believed that time in the wilderness would "ensure the survival of these qualities: an enterprising curiosity, an indefatigable spirit, tenacity in pursuit, readiness for sensible self-denial, and, above all, compassion."

The focus on values remains the core of the organization's mission today. As several Outward Bound instructors told me during my time in the field: "We don't teach for the mountains. We teach *through* the mountains."

"If you come out here, and all you do is walk the trail, eat the food, and crap in a hole, you haven't done Outward Bound," Vince said to me the day before his students got all kinds of lost. Vince was born in Indiana, worked for Toyota at its plant in Cincinnati after graduating from college, and then in his twenties dropped everything and headed for Colorado to make a different kind of life. As he sees it, part of his responsibility as an instructor is to make a love for the wild and the lessons it can bring "contagious."

He told me: "As an early Outward Bound instructor said, 'If we take you to the mountains, and you fall in love with the mountains, and you stay there—we've failed.' The point is that you'll take what you learned here and take it back out and share it with people to make the world a better place. The reason it works here is that the wilderness is unpredictable—and that unpredictability yields those outcomes of leadership and compassion and service. It's not a climate-controlled environment out here."

Big words. And, from what I witnessed, they're totally true.

Outward Bound begins with some basic backwoods ethics—Leave No Trace, Take Only Memories—and then seeks to go deeper into civic values of leadership, service, and responsibility. Poet Gary Snyder wrote that there's "an etiquette of freedom" to be found in the wild. The wilderness is also a place that can inculcate an ethos of solidarity. It teaches a kind of rugged communitarianism. The primal trials of heat, cold, wetness, dryness, hunger, thirst, discomfort, and fatigue can lead to the sort of lessons that transcend ideological divisions and political disagreements. There's something important in the fact that both well-known Colorado conservatives and well-known Colorado liberals are financial supporters of Outward Bound. It's a place where universal values are taught.

Talking with the students as we hiked, I found that most of the boys had, in fact, taken the lessons that were offered. Ade told me that back in London he's fiercely competitive and always on the go. Some time in the wilderness had forced him to slow down and appreciate the present and also to have a new appreciation for humility. "Fellowship is just as important as leadership," Ade said at one point. The solo experience had been especially instructive. "When I was on the first night of the solo, I thought I saw an animal run by, and I was seriously scared. I stayed under my tarp with a stick near me the whole time. And I came to realize that, individually, we are all rather small. But together we are much bigger. Together we can climb that mountain there. I couldn't do that alone."

The wilderness—precisely *because* it is strange and has come to seem otherworldly—forces such re-reckonings. For Nathan, a computer programmer, that meant gratitude for the beauty of the natural world: "Out here in the wilderness things are more vivid and more colorful. The world is more beautiful and more serene. You're surrounded by trees and wildlife and you look at a gigantic mountain face and you feel small, but you know that there's a connection between you and everything around here."

Jon, the cool kid–hockey player, came to a new understanding about the importance of compassion. "For me, the lesson is in thinking about how to be in someone else's shoes," he told me as we were hiking up the Piney River. "That's still not easy for me. At home—not to be a dick—if I see someone struggling, I'm like, *That sucks.* Out here you can't do that. If someone is hurt, you have to think about how they're feeling. It's like an enforced empathy. You have to feel compassion for the other person, or else you're not going to make it."

Pat arrived at another conclusion, a recognition that compassion toward wild nature is as important as compassion toward people. "Just living that way, I think it makes me more aware of, sort of, human beings' impact on the environment," Pat said. "When you live that long by the Leave No Trace code and you're with it for so long, and you come to a campsite where there is toilet paper and beer cans and chip bags strewn around, it bugs you."

Pat and I were sitting on a downed log in a grove of mostly dead lodgepole pines, and Pat motioned toward the sepulchered scene as a way of explaining the importance of showing empathy for the environment. He said, "From the very first day, we were seeing all of these dead trees. Because of human interference, and consequently climate change, we see all of these dead trees. I had expected to just see green slopes, and it was jarring. I would say it makes you more compassionate toward the environment, and the animals living in it."

In the last ten years the pine bark beetle has destroyed huge swaths of forest from Alaska to the Southern Rockies. It used to be that the beetle's population was kept in check by freezing winters. Now, climate change's milder winters have allowed the beetle's numbers to explode, and the forests have suffered. I had read about this phenomenon; I knew how the huge numbers of pine deaths had fueled Colorado's massive wildfires. Still, it was harsh to see the damage myself. The mountains above Vail should have been green, but instead they were streaked with the gray of what Pat called "dead husks." The scene looked corrosive—like a biological rust had swept through the woods, a flameless fire.

As Pat talked about how the pine beetle damage made him feel, it became clearer to me that in the Anthropocene the wilderness's lessons will be found amid transformed landscapes—ecosystems inevitably marked with civilization's fingerprints. The lodgepole and ponderosa pine forests of Colorado might not return after the beetles and the wildfires. With ever-warming temperatures the areas that were once the home of tall trees might become chaparral, or perhaps a pinyon-juniper mix like the Gila. As ecosystems shift, our mental pictures of wilderness will need to shift, too. Our idea of wilderness will probably become a bit less majestic, a bit less sublime. Yet even as we witness some awful changes, we'll have to commit to keeping some places totally unmanaged. We'll have to hold onto wildness as a touchstone for our relationship with nonhuman nature.

In short, we'll have to shrink our idea of wilderness even as we expect more from it. We'll have no choice but to settle for a physical wilderness that is something of a flawed masterpiece. And, at the same time, we will have to lean on the idea of wildness more than ever, as a gauge for measuring our impacts on the rest of creation. *Doing more with less*—it's an all-too-perfect notion for young people like Pat and Nathan and Jon, destined to live in an age of austerity.

Once again, it was André who grabbed my imagination. On the second-to-last night of the course we had to make a long slog through a steady rain to get to camp. The misdirection on Piney River—combined with frequent and lengthy snack breaks throughout the day—had put us way behind schedule. Then Brandan got some weird stomach pains at the top of the last pass, forcing us to stop, so that night was falling and a storm was approaching as we made our final push.

Instructor Nate led us through some perilous bushwhacking along a steep mountain slope, and by the time we found the Forest Service road of the front-country, the rain was coming down. We marched through the dark and the wet. The group was almost at the finish, but morale was in the mud.

We stopped for a quick breather, and some instinct—an impulse that rose out of nowhere—prompted me to ask the students to turn off their headlamps and just be still. After a minimum of grumbling, everyone shut off their lights. Someone joked how it would be cool to have night vision.

Then, as we stood there in the rain, the night *did* become visible. The shapes of the trees appeared, as did the curves of the forest terrain and the puddles reflecting the cloud glow. Silence descended. The only sound was the raindrops pattering against needle and leaf. And I heard André say in a soft voice, as much to himself as to anyone listening, "I love the nature."

The wild world revealing itself to those willing to work for it and wait for it.

If that sounds sentimental or romantic—well, it is. After all the intellectual debates and the carefully crafted arguments and the inevitable overthinking, the task of conserving, preserving, and restoring the remains of wild nature will be a labor of love. In the end, awe trumps reason. Old-fashioned wonder outweighs irony. Standing in the dark in the rain, I was reminded that to catch the spirit of the wild, the most important organ is not the ear or the eye, but the heart.

I hope it's not too obvious a metaphor to point out that we are all on a life raft together. It's about 25,000 miles in circumference, 93 million miles from the nearest star (a rather puny one, galactically speaking), with more than 7 billion mouths to feed and more passengers climbing on daily.

Navigating this new century will be far from easy. The pressures of an overheated and overcrowded world are going to force some tough choices. Should we resort to atmospheric geoengineering to counteract global warming? Should we use the power of synthetic biology to create new life forms to help feed a growing human population? Should we wedge our telecommunications into every corner of Earth? Are we willing to limit our numbers and curb our appetites so that we can share space with other creatures?

The so-called eco-pragmatism offered by those who trumpet the arrival of a garden planet will prove insufficient to answer those questions. Pragmatism is useful, no doubt, but it supplies a cramped kind of ethics. Ultimately, the idea of ecosystem services (measuring the value of life by calculating its practical worth to humans) only offers the guide of human self-interest. To grapple with the dilemmas of the Anthropocene, we'll need something more—something closer to grace. A reforged commitment to wildness can supply that. I believe we are about to discover that the wild can be like a multitool in our twenty-first-century survival kit.

For starters, big, remote wilderness will serve as a harbor for all those plants and animals that we've driven from their homes. We won't be able to save everything; the casualties will continue to mount. Yet it's essential that we keep some places free from our intentions, areas where evolution can continue to unwind without human direction. Those last resorts will be a refuge for human hope as well as haven for flora and fauna. Wildlands will keep alive the possibility of a renewal that will someday come.

The wild can also work as anchor, a counterweight to the force of industrial society. The garden has become such a potent ecological metaphor because it represents balance—a way of thinking about how to reconcile human hungers with the needs of nonhuman nature. The garden is the place where most of us make our homes; it is the middle way. But to strike that balance requires something that can counteract the huge mass of global consumer capitalism, something on the other side of the scale. Wilderness carries such weight. The wild represents the radical idea that life on Earth is not here just to suit our ends. Such a recognition can help to steady civilization even in a storm of our own making.

And, finally, the wild can act like the needle on society's moral compass. "Ego and pride—they don't serve you very well in the wilderness, I don't think," Ade said to me. That sort of humility is an all-too-rare resource these days—and it's exactly what we'll need to prevent human appetites from eating the whole planet. One day on the trail Jon told me that the experience of the wilderness, as opposed to the hockey rink, is characterized by an "unrewarded fortitude." In the sports arena, people cheer when you score a point; in the mountains, no one's even around to notice when you bag the peak. Protecting Earth's last wild places will require, above all, a steady, unheralded effort. The long march through a night rain, as it were.

The last night of an Outward Bound course is a time for a closing ceremony in which students are awarded a pin certifying that they passed the tests. Another thunderstorm had just rolled through, and the grasses were sopping wet and the path muddy as Vince led the group to a clearing where he had made a compass design out of sticks and stones. The students were given a chance to talk about what they had learned during the course and to share some appreciations for each other. After the days of bickering in the backcountry, I was surprised to hear the kids offer so much heartfelt praise to each other. Finally, Vince gave this valedictory speech:

> Tomorrow, y'all are going home. I think it's a strange dynamic to think about whether you are going *In* by going back, or if you are going *Out*. I feel more comfortable here, so when I leave, I feel like I'm going *Out*. So tomorrow you are truly outward bound. For the rest of your life you are truly outward bound.
>
> Think of all the things we might not actually know that we learned. And think about how we learn those: by living simply in a very wild place. A place where we don't see Man very often. So if you get home and maybe kind of forget what that pin stands for, or what happened out here, maybe it's time to go back to the mountains, or the forest or the rivers or the canyon. Because this is actually where we come from. And believe it or not, there is a connection. Subconsciously we are very, very connected to this place. And what we're doing out here is incredibly simple. *Out there* is complicated. But the lessons that we pull from here, from this very simple place, it's quite astonishing, in my opinion.
>
> Value the land. It is important. Even if it's just a little piece in your yard or your neighbor's yard or a city park. Everything we have comes from here. The clothes we wear. The food we eat. The lessons we learn. Don't forget our classroom.

Just then a long grumble of thunder growled overhead. Instinctively, we all looked up. The storm had moved off to the northwest, leaving the forest dripping and the sky directly above impossibly clear.

My God, the stars! Countless bursts of incandescence, gems flung across a carpet of black velvet. The Milky Way like a veil of silver mist. I tried to pick out of the Greek names for the timeless arrangements: the bears Ursa Major and Ursa Minor, the Plowman, and Virgo the virgin. From Moment Go we've been hitching our myths onto wild nature.

Then a star on the move caught my eye. It traced an exact arc across the sky, something out of sync with the stillness. A satellite. Beauty interrupted by the traffic of a million conversations.

I took a breath. Once more I wondered at the way in which Earth, even marred, remains a perfect mystery.

Acknowledgments

THIS BOOK WOULDN'T HAVE HAPPENED without Barbara Dean, who understood what I was trying to say even before I did. Barbara was one of the founding editors of Island Press more than thirty years ago, and this is to be among her last books in a career that has spanned at least 300 titles. I am humbled to be in the company of the luminaries she has edited. I can see why they came to her. Barbara is a keen reader with a knack for mixing encouragement with criticism, and she's possessed of a prescient mind. She's an editor's editor, and it was a real pleasure to work with her. Thank you so much, Barbara.

I am especially grateful to the more than ninety people (see Sources) who offered their time to talk to me for this book. A book is nothing without its sources, and I learned so much from everyone who shared their opinions with me. I hope I've been able to distill some of the wisdom of the crowd.

It's hard to imagine this book without Kristin Link's map illustrations. She did an amazing job in a small amount of time, after many other designers bowed out because they said it was too hard. Beautiful work.

Several people played key roles in putting together some of the wilderness trips chronicled here. I'm grateful to Dan Ritzman for including me on the Arctic rafting trip, Peter Steinhauser at Outward Bound for making the Colorado trip happen, and Jean and Peter Ossorio for organizing the Gila wolf-tracking expedition. Nicky Oulette provided guidance about reporting on the Pine Ridge Reservation, and she made introductions to some of the Lakota tribal members there.

A small group of advance readers provided invaluable feedback in the form of sharp-minded story memos, curt fact checking, and smart suggestions. I'm indebted to Kiera Butler, Kevin Fingerman, Nell Greenberg, and Brian Smith for making this a much better book than it otherwise would have been.

Michael Brune, Cristina Eisenberg, Peter Landres, and Milo Yellowhair read last-to-final drafts in order to double-check some of my facts and help ensure accuracy—I appreciate your eleventh-hour attention to detail.

My parents, Ron and Susan Mark, have been lifelong sources of encouragement and support. On this project, Mom (a travel agent) set up all airline flights and car rentals. Thank you!

I'm grateful to everyone who provided me with some wonderful settings in which to write. To Peter Barnes, Susan Page-Tillet, and the whole team at the Mesa Refuge in Point Reyes Station—Thank You once again. Robbie Greenberg's and Lisa Rich's home in Bellingham, Washington, has been a reliable source of inspiration and rejuvenation; good portions of three chapters were written in the Meadow, which is where it all began. Much of the thinking behind this book occurred at Sim Van Der Ryn and Francine Allen's cottage in Inverness, California. Without Nadine Oliver's little cabin on Orcas Island, I don't know how I ever would have completed the manuscript.

I'm grateful to Jay Golden, Mark Hertsgaard, Anna Lappé, Tim McKee, and Malcolm Margolin for offering professional advice and writerly wisdom about storytelling and the book business.

In addition to the indefatigable Barbara Dean, the good people at Island Press have been fantastic collaborators. Many thanks for all the support from Mike Fleming, Maureen Gately, Erin Johnson, Jamie Jennings, Julie Marshall, David Miller, and Rebecca Bright, who nailed the title.

My adventuress of a lifetime, Nell, has been consiglieri, cheerleader, and all-around BS meter. No one marks up a page better than her. I owe you one, Babes.

All mistakes, inaccuracies, and errors in judgment are my own.

Sources and Inspiration

B Y TRAINING AND HABIT I'm a journalist, not a scholar, which means I have something of an allergy to numerical notes in the text. Most of my principle sources are mentioned in the body of the book. Unless otherwise noted in the text itself or in the Notes, all quotes from individuals were told to me. A complete list of all interviewees is found below. The Notes section consists of amplifications and asides I couldn't resist, as well as substantiation of sources on issues that might be contestable or that involve recent research findings. The Bibliography lists books that have been a source of information, insight, and inspiration. Consider all of these books as suggested reading.

Interviews

In researching and reporting for this book I conducted more than ninety interviews. Most of those interviews were audio recorded; about a third of them were collected in my notebooks. Many of the people who were generous enough to share their time with me did not make it into the final manuscript, but their opinions and observations were, without exception, invaluable to my understanding of context. In the interest of more casual storytelling, in some chapters I did not include the last names of principal characters. Here is a complete list of everyone I spoke to, including their professional affiliation (whenever applicable).

Eve Ahlers

Lilian Akootchook

Peter Barnes

Sherry Barrett, US Fish and Wildlife Service

Brandan Bednarz

Jon Blust

Michael Brune, Sierra Club

Eric Brunnemann, National
Park Service

Katherine Calbert

Jess Carey

Alexia Chimenti

Chris Chimenti

C. J. Clipper, Oglala Lakota
Tribal Council

Steve Costa

Wink Crigler

Emma Doige

Jeff Dolphin, Arizona Game
and Fish Department

Maggie Dwyer, US Fish and
Wildlife Service

Trudy Eccofey

Peter Elstner, Arctic Wild

Phyllis Faber, Marin
Agricultural Land Trust

Dorothy FireCloud,
National Park Service

Matt Forkin

Brandon Gaudelli

Elena Gellert (and her
partner, Robert)

Glyn Griffin, Catron County
Commissioner

Matt Forkin

Nate Freeberg, Outward
Bound

Bruce Hamilton, Sierra Club

Heather Hardy

Mike Her Many Horses

Chuck Jacobs

Gideon James

Sarah James, Gwich'in
Steering Committee

John Jarvis, National Park
Service

Shelton Johnson, National
Park Service

Susan Johnson, US Forest
Service

Roger Kaye, US Fish and
Wildlife Service

Peter Landres, Aldo Leopold
Center

Shauna Langfield

Pat Lansdale

Karen Lone Hill, Oglala
Lakota College

Princess Luca, Gwich'in
Steering Committee

Kevin Lunny, Drakes Bay
Oyster Company

Jill Majerus, World Wildlife
Fund

Craig Mallett

Rue Mapp, Outdoor Afro

Emma Marris

Olowan Martinez

Wilmer Mesteth

Craig Miller, Defenders of
Wildlife

Paul D. Miller (aka DJ
Spooky)

Dustin Moore, Outward
Bound

Willa Moore

Jen Henrik (Seni)
Munthe-Kaas

Reed Noss, University of
Central Florida

Joe Bill Nunn

John Oakleaf, US Forest
Service

Adeyemi Ademide Olatunde

Jean Ossorio

Peter Ossorio

Peter O'Neil, Outward
Bound

Dave Parsons

Alex Patrick

Stuart Pimm, Duke
University

Cherie Pollach

Tom Poor Bear, Oglala
Lakota Tribal Council

Will Rapuano (your name
in a book, dude!)

Dan Ritzman, Sierra Club

Michael Robinson, Center
for Biological Diversity

Reed Robinson, National
Park Service

Rider Roland

Epona Rosa

Mike Roselle

Michelle Ross

Eddyr Rouki

Harold Salway–Left Heron,
Oglala Sioux Parks and
Recreation Authority

Laura Schneberger

Andre Senecharles

Curtis Temple

Amy Trainer, West Marin
Environmental Action

Nathan Van Doorn

Sim Van der Ryn

Lynx Vilden

Nicholas (Vince) Vincent,
Outward Bound

Shirley Weese-Young, Sierra
Club Foundation

Jamie Weaver

Jeff Whalen

Sylvan Willig

Allison Warden

Jessie Watson-Brown

Milo Yellowhair,
KILI Radio

Notes

Prologue

There's no such thing as bad weather . . . This idea comes courtesy of John McPhee: "With the right gear, it is a pleasure to live with the weather, to wait for sun and feel the cool of rain, to watch the sky with absorption and speculation, to guess at the meaning of succeeding events." The passage appears in the piece "The Keel of Lake Dickey" in McPhee's collection of *New Yorker* writings, *Giving Good Weight*.

Experiment with solitude . . . The phrase "experiment with solitude" is borrowed from the indefatigable Edward Abbey and appears in the essay "Come On In" from *The Journey Home: Some Words in Defense of the America West*, which I stumbled across in a Sierra Club book, *Words for the Wild*.

Unfortunately, my intrepidness . . . A gear pro-tip: since the splinter episode I have hiked Aravaipa Canyon two other times, and have come to the conclusion that the best footwear for navigating the ever-winding creek would be a pair of sturdy Keens, open and light for easy drying, but with more solid foot and toe protection.

A long, flat piece of tree . . . If an inch-long splinter seems small to you, please put a ruler to the top of your foot and measure. Then let's talk.

A peopled wilderness . . . I do not know for certain that the campsite has an Apache—much less a Mogollon or deep Paleolithic—lineage. To confirm that would require me returning to the canyon with an archaeological team.

But all the signs point to long-term human visitation. The camp sits at the fork of the main creek and a feeder stream, which makes it a natural gathering place. Its location right at the base of the canyon walls means it's high enough to avoid the biggest floods. The hearth stone's soot patina seemed deep.

Then there's this:

In January 2014, I returned to the canyon as part of the research for this book and, after a couple days of searching, finally rediscovered the spot. I spent part of an afternoon exploring the sub-canyon of the tributary stream. It was a twisting, narrow defile, the limestone there bone-white, and I felt a weird aura about the place, a haunting vibe. I clambered through a couple of S-turns in the canyon, climbed a massive deadfall, bouldered up a hump, and soon happened upon one of the most amazing places I have ever been.

A hidden waterfall was tucked in the rock. The stream poured from the heights into a deep, dark pool of blue. Over the ages the water had cut a perfect groove into the rock, so that water sluiced through a long, straight, slick chute. Maidenhair ferns and braids of moss and long lengths of green grass hung from the bowl of stone. The symmetry of the scene—the slick chute, the limestone looking as soft as flesh, the maidenhair cloaking the curves— reminded me of a woman's secret spot.

I was struck by a memory-spell of something ancient. A flash of intuition: this had been a sacred space. Perhaps a place for making trysts, a marriage location. Or maybe it was a site for initiations, the outdoor temple of a fertility cult—the water and the chute like some kind of totemic birth canal.

The campsite down below with the fire-scarred hearth wasn't the destination for whoever once came there. It was more like the antechamber.

Chapter 1

Each side made the predictable appeals to science . . . The online archives of the *Marin Independent Journal* (www.marinij.com) proved essential for tracking the ebbs and flows of the oyster controversy, as did the archives of the the *Point Reyes Light*, which are accessible only at their editorial offices, located behind the US Post Office in Inverness, CA. The first quote comes from Mark Dowie, "What's Become of the EAC?" *Point Reyes Light*, October 21, 2010. The second quote is from Joe Muller, "Doing What's Right for the Ecology," *Marin Independent Journal*, July 31, 2013.

The Oracle of Science . . . The oyster farm controversy is too byzantine to get into all of the details here. For a deep dive on the issue, see John Hart's *An Island in Time: 50 Years of Point Reyes National Seashore* (Light House Press, 2012). Another indication of how hot the issue became: in an opening Author's Note, Hart writes that the oyster controversy was so "radioactive" that he considered omitting it from his history. But that, he says, would have been like "writing about recent U.S. history with no mention of Afghanistan." Unable to find a consensus on the issue among peer reviewers, University of California Press withdrew from publishing it.

"Taliban-style zealotry" . . . Dave Mitchell, "Renew the Lease," letter to the editor, *Point Reyes Light*, November 4, 2010.

"The viciousness is beyond . . ." Carlos Porrata, "Vicious Beyond Belief," letter to the editor, *Point Reyes Light*, November 4, 2010.

"The brutalizing pressure of metropolitan civilization" . . . The line comes from one of the founding documents of The Wilderness Society, as cited in *Driven Wild*, historian Paul Sutter's complete history of how mid-twentieth-century changes, especially the automobile, spurred conservationists to defend the wild.

According to one estimate . . . This comparison comes courtesy of Roderick Frazier Nash in the latest edition of his classic, *Wilderness and the American Mind*, p. 380. Globally, about 13 percent of terrestrial area and about 3 percent of marine area are under some sort of protection as parks, preserves, or wilderness reserves, according to figures from the International Union for the Conservation of Nature and the United Nations Environment Programme.

"What right-minded environmentalist . . ."   Herb Kutchins, "Anti-Oyster Magic," letter to the editor, *Point Reyes Light*, November 4, 2010.

By the time Interior Secretary Ken Salazar . . .   Gauging public opinion on a heated topic is always tricky. The first round of National Park Service public comments on the oyster farm ran 3,600 respondents in favor of closing the oyster operation to 800 in favor of keeping the farm, but the majority of comments came from out of state. According to a 2012 online readers poll by the *Marin Independent-Journal* (often cited by oyster farm defenders, but unfortunately no longer available online), 85 percent of Marin residents supported keeping the oyster operation. The sheer number of "Save Our Drakes Bay Oyster Farm" signs that dotted the roads of West Marin well into 2014 is evidence enough that local public opinion was behind the Lunny Family. Postscript: In January 2015, after exhausting all of his legal appeals, Kevin Lunny closed the oyster farm.

"There is no ecosystem in Marin" . . .   Josh Churchman, "Seals and Human Beings," letter to the editor, *Point Reyes Light*, March 25, 2010.

Another reader argued . . .   Crawford Cooley, "The Roots of the Debate," letter to the editor, *Point Reyes Light*, December 2, 2010.

The love of the wild may be maladapted . . .   This turn of phrase comes courtesy of Stephen J. Pyne, "Green Fire Meets Red Fire: Environmental History Meets the No-Analogue Anthropocene," an essay that appears in the new anthology, *After Preservation: Saving American Nature in the Age of Humans*, ed. Ben A. Minteer and Stephen J. Pyne (Chicago: University of Chicago Press, 2015).

The Anthropocene . . .   At this point, a formal declaration that we have entered a new planetary epoch seems a matter of *when*, not *if*. Stratigraphers are mostly debating at what point in time to mark the beginning of this epoch. At the start of the Industrial Revolution? At the beginning of the Neolithic Revolution, when humans began chopping down trees to plant crops? The dawn of the nuclear era? It doesn't much matter. At the very least we now have a name for our overweening power.

"It's we who decide" . . .   Paul J. Crutzen and Christian Schwägerl, "Living in the Anthropocene: Toward a New Global Ethos," *Yale 360*, January 24, 2011. Available at http://e360.yale.edu/feature/living_in_the_anthropocene_toward_a_new_global_ethos/2363/.

The biologists' argument appeared . . .   Peter Kareiva, Michelle Marvier, and Robert Lalasz, "Conservation in the Anthropocene: Beyond Solitude and Fragility," *Breakthrough Journal* (Winter 2012). Available at http://thebreakthrough.org/index.php/journal/past-issues/issue-2/conservation-in-the-anthropocene.

"Protecting biodiversity for its own sake has not worked" . . .   Perhaps. I would point out that promoting social justice for the sake of justice hasn't entirely worked, either, nor has promoting peace for the sake of peace. But that doesn't mean we should stop trying.

The essay sparked a heated backlash . . .   See Emma Marris, "New Conservation Is an Expansion of Approaches, Not an Ethical Orientation," *Animal Conservation*, April 2014; Michael Soulé, "The 'New Conservation,'" *Conservation Biology* 27, no. 5 (2013); Michelle Marvier, "New Conservation Is True Conservation," *Conservation Biology* 28, no. 1 (2013).

The squabble swelled into a public schism . . .   See D. T. Max, "Green Is Good," *New Yorker*, May 12, 2014; Keith Kloor, "The Battle for the Soul of Conservation Science," *Issues in Science and Technology* (Winter 2015).

The animosity has become so intense . . .   Heather Tallis and Jane Lubchenco, "Working Together: A Call for Inclusive Conservation," *Nature*, November 5, 2014.

What you could call the "Nearby Nature" . . .   I am hardly the first person to have used this phrase, which has become something of a term of art in environmental circles. By way of example, see the Sierra Club's fact sheet regarding its "Our Wild America" campaign, available here: www.sierraclubfoundation.org/sites/sierraclubfoundation.org/files/Our Wild America Fact Sheet.pdf. The Sierra Club's emphasis on regional parks and close-to-home preserves is evidence enough that environmental groups are not forsaking the nearby nature and only interested as remote wilderness, as some have stated.

The word comes from the Old English *wildéor* . . .   For this breakdown of the etymology of *wild* and its connection to the idea of "will" and "autonomy," I'm indebted to environmental historian Roderick Frazier Nash, who dives into the meaning of the word at the beginning of his seminal book, *Wilderness and the American Mind*.

The wild—as a place and as a state of mind—is as close as you can get to the triggering ideal of environmentalism . . . Conservationist? Preservationist? Environmentalist? There are probably as many environmental sub-groups as there are individual environmentalists, and the taxonomy isn't all that helpful. Dave Forman, a cofounder of Earth First!, makes a distinction between "conservationists" who want to protect natural ecosystems for their own sake, and "environmentalists" who want to steward natural resources for human use. The names and affiliations shift by the decade. After a while, the discussion becomes the eco-equivalent of asking how many green angels can dance on the head of a recycled pin.

In my lifetime, humans have destroyed . . . See the World Wildlife Fund's "2014 Living Planet Report," published September 30, 2014. Abstract, with link to the full report, available at: www.worldwildlife.org/press-releases/half-of-global-wildlife-lost-says-new-wwf-report.

We need the Other . . . Some people might argue that in thinking of wild nature as the Other, we separate ourselves from it. Not necessarily. You and I are distinct, individual persons, just as we are distinct from wolves in the wild. Yet we three are still joined by shared interests—say, the basic need for clean air, clean water, and some space to roam. Solidarity relies on autonomy. We can be distinct from one another and still be connected. Another debt to acknowledge: Jack Turner's *The Abstract Wild* has one of the most thoughtful explanations for the psychological importance of the wild as Other.

Some critics of wilderness sneer that conservation today . . . For one example of a wilderness skeptic making a pejorative comparison between museums and nature preserves, see Fred Pearce, "True Nature: Revising Ideas on What Is Pristine and Wild," *Yale 360*, May 13, 2013. Available at: http://e360.yale.edu/feature/true_nature_revising_ideas_on_what_is_pristine_and_wild/2649/.

Chapter 2

There is a whole sub-genre of nature lit . . . See: Garett Reppenhagen, "An Iraq War Veteran Fights for Public Lands," *High Country News* "Writers on the Range" syndication service, January 23, 2014. See also: Matt Jenkins, "Nick Watson: Bringing the Wilderness Solution to Vets," *National Geographic*, May 2014; Jill Neimark, "The Camping Cure," Aeon.com, January 22, 2014. The last reference is to Terry Tempest Williams's modern classic, *Refuge*.

When the Whites cleared the Ahwahnechee Indians out of Yosemite . . . For the complete tragic story, see Rebecca Solnit's *Savage Dreams*.

Writes Roderick Frazier Nash . . . Nash's *Wilderness & the American Mind*, first published in 1967 and now in its fifth edition, is a must-read if you want to take the deep dive into intellectual history of the wild. I am indebted to Nash's trailblazing. The historical sections in this chapter are a distillation of his exacting work.

Muir was a major author of his day . . . For evidence of Muir's lasting influence, look no further than the coins in your pocket. The California quarter shows Muir gazing over Yosemite Valley's Half Dome, a raptor flying beneath him. The state quarters offer a fascinating glimpse into the way in which Americans' self-identity remains hitched to the wild. Seventeen quarters depict natural scenes or wildlife, more than depict what I would call pastoral scenes (four) or historical events (twelve). Even today, Americans see ourselves through the prism of wilderness myth.

No national figure embodied the era's lust for wilderness like Theodore Roosevelt . . . For the definitive story about Roosevelt's conservation activism, see historian Douglas Brinkley's *The Wilderness Warrior*. While Roosevelt's conservation ideas were progressive for his time, he was also a man of his time—a true believer in Manifest Destiny, an indefatigable imperialist, a sworn foe of the wolf and the Indian. Brinkley reminds us: "Roosevelt even considered the genocide of the Native Americans as heroic. 'The most ultimately righteous of all wars is a war with the savages, though it is apt to be also the most terrible and inhuman,' [Roosevelt] wrote."

The wilderness didn't play a role in the landscape of my childhood . . . I should say, by way of accuracy and fairness, that my parents were enthusiastic national park visitors and day-hikers. We took many road trips to natural wonders across the West. My parents also sent me to four summers of a cowboy-themed sleep-away camp in Prescott, Arizona, where I was taught to ride a horse and to start a fire with a flint. These experiences clearly helped spark in me a love of the outdoors.

At the turn of the last century, the fight over damming the Tuolumne River . . . The dispute over damming the Tuolumne River and the flooding of Hetch Hetchy Valley would be replayed in the mid-twentieth century as conservationists fought the damming of the Colorado and

Green Rivers, which would have flooded the Grand Canyon and Dinosaur National Monument. At the time Wallace Stegner wrote: "If we preserve as parks only those places that have no economic possibilities, we will have no parks." A useful reminder that every conservation battle involves a tough choice between human interests and the needs of other species, and that each act of preservation involves some sacrifice of human desire.

The term—coined by a fisheries scientist charting the ever-shrinking size of wild fish . . . See Daniel Paul, "Anecdotes and the Shifting Baseline Syndrome of Fisheries," *Trends in Ecology and Education* (October 1995). More recent research suggests that the shifting-baseline phenomemon can sometimes occur in the space of just a few years. For example, surveys of individuals in Alaska reveal that between 2004 and 2008 people's perceptions of the severity of bark beetle infestation decreased significantly even though the physical condition of the forest did not improve. That is, the more people got accustomed to the sight of dead trees, the more they thought of the situation as "normal." See Hua Qin et al., "Tracing Temporal Changes in the Human Dimensions of Forest Insect Disturbance on the Kenai Peninsula, Alaska," *Human Ecology* (February 2015).

It was the first sound of an engine I had heard in days . . . I stopped going to the Yosemite backcountry years ago because I find the constant sound of jet traffic to be insufferable. The trip with Chris, Alexia, and Michelle was the first time I had returned to Yosemite since 2002. Frankly, I was surprised that we heard no jet sound during our trek through the Grand Canyon of the Tuolumne. I can only guess that this was because the roar of the river drowned out the drone of the aircraft.

Chapter 3

From 1970 to 1997, the number of jet flights . . . The air-traffic statistics come courtesy of the Federal Aviation Administration. See www .transtats.bts.gov/Fields.asp?Table_ID=259.

Health researchers have established that noise pollution . . . European researchers have done some of the most thorough work on the consequences of noise pollution. See the European Commission's reports regarding noise pollution: http://ec.europa.eu/environment/noise/health_effects .htm.

Bernie Kraus, a musician and naturalist . . . See Maureen Nandini Mitra, "Extremely Loud: We Have Drowned Out the Natural Soundscape," *Earth Island Journal*, Spring 2013. For more on the physiological benefits of silence, see Daniel A. Gross, "This Is Your Brain on Silence," *Nautilus*, Winter 2015.

The absence of human sounds is supposed to be so profound . . . See Kathleen Dean Moore, "Silence Like Scouring Sand," *Orion*, November/December 2008.

The bustle of islands that make up the Salish Sea . . . ICYMI, the Salish Sea is the official new name for the bodies of water once known as the Puget Sound, the Strait of Georgia, and the Juan de Fuca Strait. See Isabelle Groc, "Salish Sea Change," *Canadian Geographic*, June 2011.

We humans seem predisposed to appreciate certain landscapes . . . The hypothesis that humans have an aesthetic preference for savannah-like landscapes has been pretty well established by now. See, for example: John H. Falk and John D. Balling, "Evolutionary Influence on Human Landscape Preference," *Environment and Behavior* (July 2010).

The rolling hills of Britain's Lake District, so beloved by the Romantics . . . The archetypal Romantic landscape, it's important to remember, is almost entirely a human creation. The uplands of Britain were at one time forest. Then sheep arrived and grazed it bare. The Romantics' symbol of Nature was an artifact.

I needed the patience to pay careful attention . . . The new critics of the wild sometimes say that a drawback of wilderness is that it involves no interaction; today we moderns go into the wild just to look. Perhaps what they mean to say is that the wilderness doesn't involve any interaction on human terms, any manipulation. Or maybe they've just never experienced the very real interaction of trying to keep a fire going in a snowstorm or of scrambling to set up the tent before the rainstorm hits. In any case, the observation seems off. Witnessing and listening are in themselves powerful interactions.

A century later, Annie Dillard won the Pulitzer Prize . . . It's important to remember that the heartfelt naturalism of Burroughs and Dillard didn't occur in the far reaches of the wild, but rather in the nearby nature.

It was good fun to debate which ferns were the lacy lady ferns and which were the lacy maidenhair ferns . . . In case you're curious about ferns: A lady fern *(Athyrium filix-femina)* has a long frond with many feathery, spearlike pinnules branching off the main stem. A maidenhair fern *(Adiantum pedatum)*, often five-fingered, has a more delicate main stem, often with a striking black spine. Not even in the same genus, the two ferns don't really resemble each other at all once you know what you're looking for.

Wildness preserves evolution . . . Dave Foreman explains the idea this way, along with his original italics: "*Evolution is wild.* It is wild in the deepest meaning of the word, and thus is the hallmark and highest good of wilderness."

The Endangered Species Act and the Wilderness Act were the highest legal expression of an environmental ethic that had been forming for some time . . . Once again I'm following in Nash's footsteps. His book, *The Rights of Nature: A History of Environmental Ethics*, is an essential primer on the evolution of biocentric philosophy.

A Norwegian mountaineer, Arne Naess, coined the term "deep ecology" . . . Bill Devall and George Sessions have been the most energetic popularizers of Arne Naess's philosophy. The line about ecological solidarity involving no "sharp breaks" between self and other comes from their book, *Deep Ecology: Living As If Nature Mattered*.

Much of the virtue of the true wild comes from experiencing fear . . . A hat-tip to nature writer J. B. McKinnon for his reminder about how the wilderness experience is incomplete without feeling some sense of fear. See J. B. McKinnon, "False Idyll," *Orion*, May/June 2012.

"There are places where even the Native Peoples wouldn't go, and for some reason we rush to go there" . . . The notion that Indigenous peoples of the Pacific Northwest didn't go into the mountains probably comes from early-encounter tales told to Europeans. According to *American Indians and National Parks*, the Coast Salish peoples likely were being ironic, but the European arrivals missed the point. Still, it is true that the mountains would only have been used as a trade route. The Coast Salish (which includes a wide range of tribes and nations) made their permanent habitations on the edge of beaches and the riverbanks, and saw their place in the universe as the point between the dark of the forest and the depths of the ocean.

Just a few years earlier, one of those mountain goats had killed a man from Port Townsend . . . The Park Service ended up shooting the ram. The goring episode, which happened to occur the same summer that several people died in Yosemite after being swept over the falls there, sparked a brief debate about how wild we should expect the National Parks to be. See Timothy Egan, "Goat vs. Man," *New York Times*, October 27, 2010. Though I know it's a fantasy, I like to think that somewhere deep in the Olympics the other mountain goats remember that one aggressive ram in awed terms. The Che Guevara of goats, if you will, a revolutionary who fought back. Sometimes the goats gather around to hear the story about the ram who took down the "maaaaan."

And then it was upon me—a twin propeller Chinook helicopter . . . I walked over to some other hikers' camp to confirm what I had seen. We all agreed that it must have been some kind of military training. A US Army soldier based at Joint Base Lewis-McChord in the Olympia-Tacoma area later confirmed to me that the military frequently conducts exercises over the wildlands of the Olympic Mountains.

Chapter 4

The expedition had been put together by Dan Ritzman . . . By way of disclosure, I should say that the Sierra Club paid for a portion of my trip to Alaska. Everyone else on the trip was fully paid for. But in the interest of maintaining journalistic independence, I insisted that I pay for my trip. At the same time, I didn't want to have to pay full freight—$4,500 to the outfitter alone. So we hatched a gentleman's agreement and went Dutch.

The science is unequivocal: the planet is warming . . . I am not going to belabor the point about anthropogenic climate change, and I'm going to assume that if you bought this book you acknowledge the overwhelming evidence that humans are driving global warming. But in case you have any doubts, dive into the assessment reports from the International Panel on Climate Change, available here: www.ipcc.ch/publications_and_data/publications_and_data_reports.shtml.

"It don't get that cold no more in the winter time," Gideon said . . . I would hear many of the same eyewitness reports a week later when we flew out of Kaktovik, an Inupiaq community on the shores of the Arctic

Ocean. An Inupiaq elder named Marianne told me, "The tundra is melting. The small streams—they're now big ravines, because the land is eroding. There are salmonberries right here. I saw them behind the fence over there. We never used to have salmonberries."

It's a patent injustice . . . Native communities in the Arctic aren't unfamiliar with this kind of ecological injustice. For decades, public health officials have known that the Inupiaq (aka Eskimos) and other Alaskan Natives are exposed to higher amounts of persistent organic pollutants as chemicals drift from the lower latitudes, settle to the surface due to the polar region's colder air, and then enter the food web, where they bioaccumulate up to humans. See this US EPA research review fact sheet for more: http://epa .gov/ncer/tribalresearch/publications/tribal_research_flyer052510.pdf.

A writer for the *Anchorage Daily News* summed up the takeaway from his death . . . I nabbed this *Anchorage Daily News* quote from a magazine article about McCandless and the film *Into the Wild*. See Matthew Power, "The Cult of Chris McCandless," *Men's Journal*, September 2009.

The campaign waged in the 1950s to protect the region . . . For the most complete accounting of the refuge's relationship to the broader wilderness preservation movement, I recommend Roger Kaye's thorough book, *The Last Great Wilderness.*

When you hear people talk about oil drilling in "An-Whar" . . . As you might notice, I refuse to use the term ANWR, which has become shorthand for the place during the long-running battles over oil drilling there. ANWR is the name preferred by the oil industry, and for calculated reasons. An acronym makes a place anonymous, interchangeable, and without personality. So I insist on using "Arctic Refuge," which better communicates the essence of the place.

A few of us tried to make sense of what we had experienced . . . For the sake of narrative fluidity, I have compressed several conversations into one. Every line that appears in quotes was spoken to me directly and recorded at the time.

Recent studies had revealed . . . For more on the competition between red foxes and Arctic foxes, see the authoritative and aptly named website: http://climatechangeandtharcticfox.weebly.com/.

In the foreword to a 2010 book . . . See William C. Tweed's *Uncertain Path*. In a June 2014 interview with Director Jarvis, I followed up with him on this point. Jarvis told me: "Now, at some point we are going to be challenged with issues like assisted migration. You know, we've got Isle Royale [Michigan] right now with a very, very small population of wolves. I have an old buddy who is a biologist who's said, "Hey, Jon, when are you ready to put the sprinkler system on the Giant Sequoias?" [laughs] Because, the Giant Sequoias are not going to migrate, right? Nor are Joshua trees. I mean Joshua trees *will*, but pretty slowly. So if you've got a climate that's driving these iconic species out of their present environment, we need the analytical tools to predict what's the next place that Giant Sequoias might persist for 3,000 to 5,000 years. Is that in the Southern Cascades? I don't know that yet. But, we're going to have to face those kinds of issues."

We now routinely supervise ecosystems that are otherwise undeveloped . . . See Eric Wagner, "The Last Stand," *Earth Island Journal*, Summer 2011; Jordan Fisher Smith, "The Wilderness Paradox," *Orion*, September/ October 2014; Douglas Fischer, "High in Yellowstone, a Foundational Tree Falters," *Daily Climate*, October 8, 2014; Christopher Solomon, "The Wilderness Act Is Facing a Midlife Crisis" *New York Times*, July 5, 2014; Ken Belson, "Arizona Enlists a Beetle in Its Campaign for Water," *New York Times*, July 14, 2014. Many other examples of interventions in wilderness can be found in the book *Beyond Naturalness*.

I had the chance to talk through some of these issues with Roger Kaye . . . Kaye's views are best summed up in this article: Roger Kaye, "What Future for the Wildness of Wilderness in the Anthropocene?" *Alaska Park Science* 13, no. 1. Available at www.nps.gov/akso/nature/science/ak_park_science/PDF/Vol13-1/APS_Vol13-Issue1-40-45-Kaye.pdf.

That's a difficult idea for many conservation biologists to accept . . . For more on this raucous intellectual debate, see Forum on Managing the Wild, *Frontiers in Ecology and the Environment*, available at: http://leopold.wilderness.net/pubs/531_1.pdf.

"The essential quality of the wilderness is its wildness." . . . For more on this point, including the intentions of the authors of the Wilderness Act, see Douglas W. Scott, "'Untrammeled,' 'Wilderness Character,' and the Challenges of Wilderness Preservation," *Wild Earth*, Fall/Winter 2001–2.

If the wild is going to remain meaningful, we'll have to commit to leaving our hands off, no matter the consequences . . . I would point out that there's a big difference between removing plastic trash from a remote beach and remodeling entire ecosystems to conform to our idea of "natural."

Chapter 5

The 107 families who lived in the area . . . Most historians offer the vague number of 100 families. This exact figure comes from Chuck Jacobs, whose family was among those displaced by the bombing. Curtis Temple added the details about the first and second warnings, and windows being blown out.

Somewhere beyond that was Stronghold Table . . . As a *wasicu*, I want to make sure I am perfectly clear: I had no intention of entering Stronghold Table unaccompanied by a Lakota. I was well to the east of there, not more than a couple of miles from the White River Visitor Center.

Despite leftover place names . . . Twenty-five—*half*—of US state names are derived from Native words (if you include New Mexico, from the original *Mexica*). Let them roll off your tongue with an another language in mind—*Alabama, Massachusetts, Mississippi, Nebraska, Utah.* . . . A list, with etymology, is available at: www.native-languages.org/state-names.htm.

"America's Best Idea" . . . For more on the Park Service's history with American Indians, see *American Indians & National Parks*. Unfortunately, the dark side of conservation was exported to other nations along with the ideal's noble aspirations. From Africa to Latin America to Asia, indigenous peoples have been displaced from their ancestral homes to make way for "nature preserves," as investigative journalist Mark Dowie uncovered in his book, *Conservation Refugees* (MIT, 2009). Thankfully, that is now changing. At the World's Park Congress in 2014, conservation organizations and national governments made new commitments to protecting indigenous communities' subsistence rights.

Temple is one of the biggest Indian ranchers . . . Language is a minefield. Maybe a more politically correct term would be "Native American," but I've never met a Native who called themselves that outside of an academic setting. I have heard "Native" and "Indigenous" and, most commonly,

"Indian," so that's what I'm going with. For the most part, I will strive to speak of specific nations, a word I prefer to "tribe," as it signifies sovereignty. Though my copy editor insists it is not standard style, I like to capitalize "Indigenous" when referring to a whole people. We capitalize "African" and "Asian," after all, and it seems to me that Indigenous is equally a geography, if one of time.

In December 2013, the tribal government and the National Park Service announced they were canceling grazing leases in the Badlands' Southern Unit . . . Readers eager to parse for themselves the competing claims about the National Park Service's intentions for the Southern Unit handover should read the 2012 General Management Plan for the area, available here: www.parkplanning.nps.gov/badl. For details on the original tribal ordinance, see Andrea J. Cook, "Oglala Sioux Tribe Evicting Tribal Ranchers to Make Way for Bison Park," *Rapid City Journal,* December 5, 2013.

Reintroducing a herd of up to a thousand bison to the area . . . buffalo and cattle don't mix . . . Another language kerfuffle. *Bison bison* is the accurate name, and to call the iconic beast of North America a buffalo is, really, to insult it. There's no comparing a bison to its flaccid Old World cousins. But I find most everyone says "buffalo" except biologists. I use both "bison" and "buffalo," as it suits my purpose.

Temple's ex-wife, Tammy . . . "more hard" . . . I refuse to use "[*sic*]" in the text when recording nonstandard English. It seems rude, especially given the fact that Tammy's English is worlds better than my Lakota.

And the bison reintroduction had dribbled out into the media . . . Andrea Cook at the *Rapid City Journal* has done the most thorough month-to-month reporting on the issue. Freelance journalist Brendan Borrell has written several stories about the Badlands saga for national publications; find them at www.brendanborrell.com. See also Juliet Eilperin, "In the Badlands, a Tribe Helps Buffalo Make a Comeback," *Washington Post,* June 23, 2013.

The dire conditions on Pine Ridge . . . Social indicators come from the US Census Bureau. See http://quickfacts.census.gov/qfd/states/46/46113.html. Recently, the reservation has suffered an epidemic of teen suicides. See Julie Bosman, "Pine Ridge Reservation Struggles with Suicides among Its Youth," *New York Times,* May 1, 2015.

The Lakota are still very much fighting . . . In March 2012, Lakota activists, including Oglala tribal vice president Tom Poor Bear, blocked a "heavy haul" of oil equipment headed for the tar sands that was crossing tribal land, leading to arrests. Lakota, including Deborah White Plume (whose relation to Percy I could never figure out), were very active in the Spring 2014 "Cowboys and Indians Alliance" activities in Washington, DC, against the Keystone XL pipeline. In November 2014, the Rosebud Sioux tribal council declared that construction of the pipeline would be considered an "act of war."

It's not hard to imagine how people could have wiped out a range of species . . . Such megafuana extinctions have on occasion been far less "innocent," as people watched the disaster unfold in human time. The best example is the Maori's destruction of the giant, flightless moa. In a generation or two, people ate themselves out of a food source.

The estimates of pre-Columbian human population of North America are fiercely contested . . . Conservation activist Dave Foreman makes a good point on this issue. Let's say, for the sake of argument, that there were, oh, 10 million people living north of the Rio Grande in 1491 (a number at the very highest edge of credible estimates). That would mean a population less than 3 percent the size of the combined US-Canadian population today. Most of the continent would have been only lightly settled by humans.

Nearly two years after the oyster controversy in Point Reyes was all but settled, a writer in the local newspaper . . . See Chet Seligman, "The Romantic Wilderness," *Point Reyes Light*, September 25, 2014.

Cynical "gotcha" . . . See Jonah Goldberg, "Nature Today Is Anything but 'Natural,'" *National Review* Online, April 20, 2014. Available at www.nationalreview.com/article/376815/nature-today-anything-natural-jonah-goldberg. Thanks to Goldberg's syndication, this essay was reprinted in newspapers across the country, including the *Los Angeles Times*, and popped up in my Google Alerts more than a dozen times.

It has been well document how the American Indians' sudden injection . . . Sheppard Krech III makes this point most convincingly in *The Ecological Indian*. The beaver, for example, went from an animal of middling interest to some kind of instant jackpot.

7,200 foot Harney Peak . . . Classic Americana: General William S. Harney was a notorious Indian killer who committed a My Lai–like massacre against the Brulé Sioux in 1854.

"The high and lonely center of the earth" . . . As the Indian historian Vine Deloria Jr. notes in the foreword to my copy of *Black Elk Speaks*: "Present debates center on the question of Niehardt's literary intrusions into Black Elk's system of beliefs." I agree with Deloria's conclusion on this point: "Can it matter?"

Under the Fort Laramie Treaty of 1868 . . . The Lakota claims over the Black Hills remain such a big deal today because they were such a big deal when the Fort Laramie Treaty was signed. After the carnage of the Civil War, the American public was battle-weary. Quakers and other reformers had turned their attention from the abolition of slavery to the mistreatment of Indians. Newly elected President Ulysses S. Grant was eager to appear a peacemaker, and the treaty was the centerpiece of his foreign policy. In 1870, Red Cloud and other Lakota leaders made a celebrity tour of the East, including a state dinner at the White House and a standing-room-only speech at New York's Cooper Union. The eventual breach of the treaty took place in plain sight of the American public.

Every person I spoke with repeated some version of a saying ascribed to Crazy Horse . . . While Crazy Horse is usually remembered as a cunning and courageous warrior, he was much more than that to his people. He was a *mystic*-warrior, a man whose leadership came from his reputation for spirituality. Wiry and intense, with wavy hair and fair skin that hinted at a French trapper's blood somewhere in his past, Crazy Horse was by birth and instinct an outsider. Historians' accounts describe him as "taciturn," exuding a "natural melancholia" and an "ethereal quality." He was usually silent at council fires, "a master of the sidelong glance." It was generally assumed that he could perform magic.

Crazy Horse was instinctively wild in the sense that he was untamable ("rebellious" is one of the definitions offered for *wild* in the OED). He was also wild at heart. "For most of his life he avoided not only white people, he avoided people, spending many days alone on the prairies, dreaming, drifting, hunting," Larry McMurtry writes in his short biography of the man. "There was something of the hermit in him." Crazy Horse was the kind of person who, to borrow from the language of the Wilderness Act, required ample "opportunities for solitude." He was most at home in the backcountry.

"We say all land is sacred," Milo Yellowhair told me . . . Yellowhair had one of the most sophisticated strategies for reclaiming the Black Hills that I heard while on the reservation. He suggests that a president, say Obama, could issue an executive order requiring that the US Forest Service's Black Hills unit have a preferential hiring practice for Lakota. Once Lakota were trained up in Forest Service procedures, the USFS could enter a co-management agreement with the tribe, assured that the land would continue to be managed in accordance with federal standards. Eventually, in a generation, say, there would be enough trust to transition the entire 1.25 million acre federal holding to tribal control, leaving private properties in the Hills untouched.

Bear's Butte and Bear's Lodge and the peaks of the Black Hills were places where the Lakota could explore the mysteries of existence . . . Some whites have complained that, since the Lakota only entered the area in the late 1700s, they can't really claim the Black Hills as sacred. Nonsense! Mount Rushmore is just barely 120 years old, and millions of Americans consider it a national shrine. There's no sell-by date on sacredness. It can be inherited, attached, or conjured. All that matters is that it's strongly felt.

Chapter 6

Some of the wolves in the Lamar Valley packs have been photographed and filmed to death—literally . . . The removal of the gray wolf from the endangered species list and the resumption of wolf hunting on the boundaries of Yellowstone National Park has not only led to the death of wolves habituated to being viewed by humans, but has also contributed to a decrease in wolf sightings among park tourists as the wolves have become newly cautious. For further details, see Cristina Eisenberg's *The Carnivore Way*.

"I got a gut shot on it," Hardy told me . . . I have had a difficult time confirming Heather Hardy's wolf-shooting tale. She said the episode "was in the papers everywhere. I almost went to prison over it, until they saw that my horse was attacked." I could not, however, locate any media reports about the incident. The closest I came were several news reports from September 2008 about "an anonymous Cruzville mother, who said her animals have been attacked by wolves over the past four years." Despite a lack of corroboration, I have decided to keep the story. Even if the tale is false or embellished, it is revealing about the intensity of anti-wolf sentiment in the area.

A wolf can travel forty miles in day . . . A Finnish biologist once reported a pack that moved 125 miles in the course of the day, as Barry Lopez recounts in *Of Wolves and Men*. In 2012 and 2013, a gray wolf dispersing out of southern Oregon and into northern Oregon, known as OR7, grabbed international headlines as it covered thousands of miles in the course of looking for a mate. According to the California Department of Fish and Wildlife, the animal traveled an average of fifteen miles per day. In the fall of 2014, a female wolf was spotted on the North Rim of the Grand Canyon. DNA tests of the wolf's scat determined that the wolf dispersed all the way from the Northern Rockies. In another example of the antipathy toward wolves, the animal was shot and killed in Southern Utah.

Interior Secretary Bruce Babbitt—a former Arizona governor— participated in the release of two female wolves . . . There are various versions of the Babbitt quote floating around the Internet. The quote I have used here combines the two most reputable sources I could locate. See Frank Clifford, "Wolves Get Help in Battling Enemies," *Los Angeles Times*, November 17, 1998; "Babbitt Says Wolves 'to Stay' in Arizona," Associated Press, November 17, 1998.

Between 2004 and 2013, only eleven wolves were sent into the wild . . . For statistics on wolf releases, translocations, removals, and poaching, visit the US Fish and Wildlife Service's page at www.fws.gov /Southwest/es/mexicanwolf/MWPS.cfm.

At least eight wolves died in the course of such operations . . . There is dispute about the number of fatalities that have occurred in the course of translocations or removals. Michael Robinson says the number is nineteen; Jean Ossorio says it's fourteen; the USFWS says it's eight.

While many other wolf populations in the United States have been removed from the endangered species list due to intense lobbying from the livestock and hunting industries . . . For the detailed story on how Congress removed a single species from endangered species listing for the first time in the ESA's history, see James William Gibson, "Cry Wolf," *Earth Island Journal*, Summer 2011. In the years since Congress removed most wolf populations from the endangered species list, federal judges have over-ruled that decision and have returned some (but not all) wolf populations to ESA protection. As of May 2015, some state and federal legislators continue to push various legislative proposals to again revoke ESA protection for wolf populations. The wolf remains the most polarizing animal in America.

Under a new USFWS rule finalized in January 2015 . . . See www .fws.gov/southwest/es/mexicanwolf/pdf/Mx_wolf_10j_final_rule_to _OFR.pdf.

This compromise approach pleased no one . . . See Tony Davis, "Ranchers, Environmentalists Alike Don't Like Mexican Wolf Plan," *Arizona Daily Star*, November 26, 2014.

The Mexican gray wolf may have been given a somewhat larger box in which to roam—but it's still a box . . . The intensive tracking and manhandling of the lobo has continued even after the USFWS rule-change. Over the course of five days in February 2015, wildlife agents cap-tured and then "processed" and collared or returned to captivity eleven ani-mals—10 percent of the known population. See the Arizona Fish and Game Department's "Mexican Wolf Reintroduction Project Monthly Update, February 2015," www.azgfd.gov/w_c/wolf/documents/02282015_MW _monthlyupdate.pdf.

One day he and some other foresters were "eating lunch on a high rimrock" . . . No one knows for certain the actual site of Leopold's now-iconic "fierce green fire" anecdote. Some people claim that it's inside the Gila Wilderness, perhaps on the Middle Fork of the Gila beneath Loco Mountain and Aeroplane Mesa. In an article in the Fall 2012 edition of *Forest History Today*, "Searching for Aldo Leopold's Green Fire," environmen-tal historian Susan Flader makes a convincing case that it actually occurred in Arizona's Blue Range Mountains, above the Black River. See http:// foresthistory.org/Publications/FHT/FHTFall2012/Flader_AldoLeopold .pdf.

According to Parsons—whose arguments are backed by several peer-reviewed studies . . . For details about the most up-to-date scien-tific findings regarding a healthy and stable Mexican gray wolf population, see Carlos Carroll et al., "Developing Metapopulation Connectivity Criteria from Genetic and Habitat Data to Recover the Endangered Mexican Wolf," *Conservation Biology* (May 2013); and a letter from eminent conservation biologists to the USFWS, June 20, 2012 and available at: www.conbio.org/ images/content_policy/2012-6-20_ASM_SCB_SER_Mexican_wolf_ letter.pdf.

Biologists sometimes describe apex predators' influence on the landscape as "the ecology of fear" . . . Wildlife biologist John W. Laundré is usually credited for coining this evocative phrase. See John W. Laundré et al., "The Ecology of Fear: Optimal Foraging, Game Theory, and Trophic Interactions," *Journal of Mammology* (May 1999).

Losing an animal that you've raised from the day it was born is a blow to the heart . . . Wildlife advocates sometimes blithely dismiss ranchers' claims that they suffer emotional hardship from predator loss. *How can you mourn an animal that you're only raising to kill?* the argument goes. This seems to me a failure of empathy on the part of environmentalists. I'm only a mere vegetable grower, but as a market gardener I also know something about the pain of predation. Over the years I've seen rows of potatoes, broccoli, and strawberries destroyed by gophers. There is a real sense of loss, anger and, yes, desire for revenge. From having worked on a friend's ranch (Animal Welfare Approved, no less), I know that such feelings are only compounded when one is raising mammals.

Many people who live in the Gila are convinced that the wolves don't just pose a threat to ranchers' livelihood, but are also a risk to human life . . . The persistent fear that wolves will attack people is completely unhinged from fact. During the last fifty years, no one in North America has been killed by a non-rabid wolf, and even attacks by rabies-infected wolves are exceedingly rare. In comparison, since 1990, bears have killed fifty-nine people and cougars have killed eleven in North America. See John Linnell et al., "The Fear of Wolves: A Review of Wolf Attacks on Humans," *Large Carnivore Initiative for Europe*, January 2002; Mark E. McNay, "A Case History of Wolf-Human Encounters in Alaska and Canada," Alaska Department of Fish and Game, Wildlife Technical Bulletin 13, 2002; and finally Ian K. Kullgren, "Department of Fish and Wildlife Says There Have Been No Wolf-Related Deaths in the Rockies," Politifact, December 16, 2011.

Many residents of the Gila are convinced that the wolf reintroduction is a government conspiracy . . . For evidence about how the opposition to wolf reintroduction has dovetailed with the broader movement against federal authority, check out a short film called "Wolves in Government Clothing" produced and written by David Spady. Spady is the California director of Americans for Prosperity, the Koch brothers–funded organization that has become a hub for anti-government and Tea Party activists. See http://wolvesingovernmentclothing.com/.

The wolf is actually antagonistic to our interests . . . That is, as long as humans remain carnivores. If the whole society went vegan—well then, problem solved. Ain't gonna happen. I, for one, intend to remain a conscientious carnivore, and so at some level will continue to have a conflict of interests with the wolves.

Kerrie Romero, a representative with a hunting organization . . . This quote comes from my notes taken during a USFWS public comment meeting on the Mexican gray wolf that was held on November 20, 2013, in Albuquerque.

The wolf makes us ask whether we're willing to share space on this planet . . . It's important to note that there are some impressive examples of ranchers making an effort to coexist with large predators. In Alberta and Montana, ranchers have worked diligently to reduce conflicts with grizzly bears. See Ben Goldfarb, "Home on the Range," *Earth Island Journal*, Spring 2014. Additional details at http://blackfootchallenge.org/Articles/. And in the Southwest, Wink Crigler and Craig Miller both belong to a coalition called The Coexistence Council, which is trying to find creative ways to reduce wolf predation. See www.coexistencecouncil.org/home.html.

The question is made more difficult by the diminishing size of our world . . . Space apart from human populations is not necessarily a requirement of human–large carnivore coexistence. A recent paper concludes that the populations of large carnivores—including bear, wolf, and wolverine—are higher in Europe (with its much more concentrated human populations) than in the continental United States. See Guillaume Chapron et al., "Recovery of Large Carnivores in Europe's Modern Human-Dominated Landscapes," *Science*, December 2014.

Chapter 7

"After four months preparing, we go time-traveling," Lynx likes to say . . . This quote comes from the Canal Plus television documentary film that was produced by Eric Valli in 2013 and which has aired in France and Quebec. It's titled, *Lynx: Une Femme hors du Temps*—in English, "A Woman Outside of Time." One night at the camp, Lynx screened a special English-language version for us.

Knowles was later accused of being a fraud . . . For more on this incredible story, see Jim Motavalli, *Naked in the Woods: Joseph Knowles and the Legacy of Frontier Fakery* (New York: Da Capo Press, 2008). When it comes to today's survivalist television shows, we're all in on the joke (look, it's Seth Rogen and James Franco naked in the woods!), so there's no such thing as "fakery." The very presence of the camera signals that the survivalists aren't in real trouble.

It's an underground movement of modern-day nomads and hunter-gatherers . . . Photographer Adrian Chesser has done some of the most evocative reporting on the subculture of contemporary primitivists. See Adrian Chesser, *The Return* (Hillsborough, NC: Daylight Books, 2014). In April 2015 I deepened my understanding of the motives behind the neo-primitivist movement by attending part of the week-long Buckeye Gathering in the foothills of the Sierra Nevada.

But here were these rosy-cheeked and bright-eyed women . . . Eventually I realized that the wildlings all seemed to be glowing from within because they were still experiencing some kind of natural high after doing the Project. Their inner glow reminded me of how some people look when they return from Burning Man for the first time. I could tell that the wildlings were still on another planet.

Every night after dinner we gathered for songs or stories . . . The Living Wild School steers very close to appropriating Native American culture, what with its buckskin clothing and the telling of Indian tales. Lynx is aware of this, and she has what I think is a convincing response: at some point, *every* human culture engaged in hide tanning and storytelling. No one has a monopoly on bow hunting or starting a fire with a bow drill.

Lynx uses the fire circle like a portal to the past . . . New research confirms that the conversations that occur at night around a fire circle are a uniquely powerful form of culture creation and transmission. After spending 174 days living with the Ju|'hoan (!Kung) Bushmen of Botswana and Namibia, anthropologist Polly Wiessner concluded that the conversations that take place around a fire at night center on singing, dancing, spirituality, and "enthralling stories," and that such interactions through the ages likely shaped entire cultures. See Polly Wiessner, "Embers of Society: Firelight Talk among the Ju|'hoansi Bushmen," *Proceedings of the National Academies of Sciences* 111, no. 39 (September 2014): 14013–14.

The wild does, indeed, resist definition . . . For one of the more thoughtful discussions on how the wild eludes words, see Brooke Williams, "A Wild That Leaves Us Speechless," *Earth Island Journal,* Autumn 2014.

Exhibit A: Google is busy making plans . . . I have written a couple of magazine articles about the new high-tech intrusions into the wilderness, and some of my earlier thoughts are repeated here. See Jason Mark, "Wifi in the Woods," *Atlantic* Online (atlantic.com), August 10, 2014; also Jason Mark, "Where the Wild Things Are," *American Prospect* Online (prospect.com), April 14, 2014.

Some of the smartest minds in America are hard at work making them reality . . . See Sam Frank, "Come with Us If You Want to Live: Among the Apocalyptic Libertarians of Silicon Valley," *Harper's,* January 2015.

Somewhere in the kernel of our consciousness we've always known . . . I'm very grateful to my editor for turning me on to the work of environmental philosopher Paul Shepard. For more on the primal relationship to nature, see his book, *The Tender Carnivore & the Sacred Game.*

How else to explain the many cautionary tales—Adam and Eve's fall from grace and Prometheus's heinous crime and the sadness embedded in the Epic of Gilgamesh . . . For a detailed treatment of what our ancient myths reveal about our long-lost relationship to wildness, I recommend the early chapters of Max Oeschlaeger's *The Idea of Wilderness.* Oelschlager points out that the Hebrew *Edhen* (Eden) is variously translated as "paradise," "plain," and "hunting ground." The original heaven on earth was a hunter's paradise.

Even a committed agrarian like Wes Jackson . . . Jackson made this point (I'm sure not for the first time) during an April 5, 2013, talk at the David Brower Center in Berkeley, California, that I attended. A video is available at www.browercenter.org/programs/wes-jackson.

Recent plant science confirms the insights of ancient myths . . . See Jo Robinson, "Breeding the Nutrition Out of Our Food," *New York Times*, May 25, 2013. It's also important to note that while we are so proud of our domestications, they are in one sense relatively narrow. We've domesticated just twenty-six of the hundreds of land mammals on Earth. Most beings resist our efforts at taming, or else we have no use for them.

There are no straight lines in nature . . . I'll admit that this common bit of rhetoric is more poetically true than precisely true. The slabs of ice on the Aichilik looked plumb straight to me, as do the hexagon cells of a honeycomb or the lines of rock at Devil's Tower. An apple falling from a tree doesn't take a serpentine route; it falls straight downward.

No man could force another man to do something against his will . . . It's important to note that pre-Columbian cultures of political equality didn't extend to women, who were often considered second-class citizens (at best).

Many scholars agree that this model of political equality was among the most important trades of the Columbian Exchange . . . Charles Mann makes the point emphatically, and eloquently, in a coda to his book, *1491*.

That anarcho-wildman Edward Abbey made the point forcefully . . . I'm also reminded of a line from the poet Gary Snyder: "In a fixed universe, there would be no freedom."

"Probably the population of the United States can't live this way" . . . Archaeologists disagree about the exact number of humans who lived on Earth before the invention of agriculture, but there is a general consensus that the number was in the mere tens of millions—say, the size of a single Asian megacity today.

At the same time, we'll have to further intensify our agriculture and grow more food on less land . . . As a committed organic farmer, I want to make perfectly clear what I mean by this. *Intensification does not necessarily mean industrialization.* Studies in agroecology by, among others, the UN Conference on Trade and Development and the UN Environment Programme, have shown than organic farming methods can match or exceed industrial agriculture yields. The trick to doing so, however, is to increase total inputs of human labor. Yet another contradiction: even as we have to urbanize we probably need to increase the number of humans engaged in agriculture, at least in the industrialized world.

Epilogue

Seventy percent of those who regularly engage in outdoor recreation are white . . . See Ryan Kearny, "White People Love Hiking. Minorities Don't. Here's Why." *New Republic*, September 6, 2013. Complete statistics can be found in the Outdoor Foundation's 2013 "Outdoor Participation Report," available at www.outdoorfoundation.org/pdf /ResearchParticipation2013.pdf.

"I find that almost no one I know who is forty or younger goes backpacking" . . . See Christopher Ketcham, "The Death of Backpacking," *High Country News*, July 21, 2014.

This being an all-boys group, there were also fart jokes . . . My only regret from my time shadowing the Outward Bound group is that I wasn't in a group with some gender diversity. The Outward Bound instructors informed me that, while most courses are mixed gender, participation remains disproportionately male.

Both well-known Colorado conservatives and well-known Colorado liberals are financial supporters of the Outward Bound School . . . I was told this by Peter O'Neil, the executive director of Colorado Outward Bound School.

Most of the boys had, in fact, taken the lessons that were offered . . . I came to my own lessons while in the Gore Range. It was difficult for me not to tell the boys they were heading in the wrong direction when they got lost. A lesson, I guess, in restraint and patience.

Pragmatism is useful, no doubt, but it's a cramped kind of ethics . . . Among the dozens of different essays in defense of wildness that have been published amid "the battle for the soul of conservation," few equal Brandon Keim's forceful, thoughtful piece, "Earth Is Not a Garden," published in *Aeon*, September 18, 2014.

Bibliography

Abbey, Edward. *Desert Solitaire: A Season in the Wilderness.* New York: Random House, 1971 edition.

Ackerman, Diane. *The Human Age: The World Shaped by Us.* New York: W. W. Norton, 2014.

Anderson, Alun. *After the Ice: Life, Death, and Geopolitics in the New Arctic.* New York: Smithsonian Books, 2009.

Banerjee, Subhankar, ed. *Arctic Voices: Resistance at the Tipping Point.* New York: Steven Stories, 2012.

Bass, Rick. *The Ninemile Wolves.* New York: Mariner, 2003 edition.

Berry, Wendell. *The Unsettling of America: Culture & Agriculture.* San Francisco: Sierra Club Books, 1977.

———. *Bringing It to the Table: On Farming and Food.* Berkeley, CA: Counterpoint, 2009.

Brinkley, Douglas. *The Wilderness Warrior: Theodore Roosevelt and the Crusade for America.* New York: Harper Perennial 2009.

Brower, Kenneth. *Hetch Hetchy: Undoing a Great American Mistake.* Berkeley, CA: Heyday, 2013.

Brown, David E., and Neil B. Carmony, eds. *Aldo Leopold's Southwest.* Albuquerque, NM: University of New Mexico Press, 1990.

Brown, Dee. *Bury My Heart at Wounded Knee: An Indian History of the American West.* New York: Henry Holt and Company, 1970.

Buechel, Eugene, and Paul Manhart, eds. *Lakota Dictionary: Lakota-English / English-Lakota.* Lincoln: University of Nebraska Press, 2002 edition.

Burdick, Alan. *Out of Eden: An Odyssey of Ecological Invasion.* New York: Farrar, Straus and Giroux, 2005.

Callicott, J. Baird, and Michael P. Nelson, eds. *The Great Wilderness Debate: An Expansive Collection of Writings Defining Wilderness from John Muir to Gary Snyder.* Athens, GA: University of Georgia Press, 1998.

Coates, Peter. *Nature: Western Attitudes Since Ancient Times.* Berkeley, CA: University of California Press, 1998.

Cole, David N., and Laurie Yung, eds. *Beyond Naturalness: Rethinking Parks and Wilderness Stewardship in an Era of Climate Change.* Washington, DC: Island Press, 2010.

Cronon, William, ed. *Uncommon Ground: Rethinking the Human Place in Nature.* New York: W. W. Norton, 1996.

Cullinan, Cormac. *Wild Law: A Manifesto for Earth Justice.* White River Junction, VT: Chelsea Green Publishing, 2011.

DeMallie, Raymond J., and Douglas R. Parks. *Sioux Indian Religion.* Norman, OK: University of Oklahoma Press, 1987.

DeMallie, Raymond J., and Elaine A. Jahner, eds. *Lakota Belief and Ritual.* Lincoln, NE: Bison Books, 1991.

Devall, Bill, and George Sessions. *Deep Ecology: Living as if Nature Mattered.* Salt Lake City, UT: Gibbs Smith, 2007.

DeVoto, Bernard, ed. *The Journals of Lewis and Clark.* Boston: Mariner, 1997 edition.

Diamond, Jared. *Guns, Germs, and Steel: The Fates of Human Societies.* New York: W. W. Norton, 1997.

Dillard, Annie. *Pilgrim at Tinker Creek.* New York: Perennial Classics, 1998 edition.

Douglas, William O. *My Wilderness: The Pacific Northwest.* Garden City, NJ: Doubleday, 1960.

————. *Of Mountains and Men*. Guilford, CT: Lyons Press, 2001 edition.

Drury, Bob, and Tom Clavin. *The Heart of Everything That Is: The Untold Story of Red Cloud, an American Legend*. New York: Simon & Schuster, 2013.

Dunbar Ortiz, Roxanne, ed. *The Great Sioux Nation: Sitting in Judgment on America*. Lincoln, NE: Bison Books, 2013 edition.

Eisenberg, Cristina. *The Carnivore Way: Coexisting with and Conserving North America's Predators*. Washington, DC: Island Press, 2014.

Foreman, Dave. *Rewilding North America: A Vision for Conservation in the 21st Century*. Washington, DC: Island Press, 2004.

Frazier, Ian. *On the Rez*. New York: Picador/Farrar, Straus and Giroux, 2000.

Hart, John. *An Island in Time: 50 Years of Points Reyes National Seashore*. Mill Valley, CA: Lighthouse Press, 2012.

Haupt, Lyanda Lynn. *Crow Planet: Essential Wisdom from the Urban Wilderness*. New York: Little, Brown, 2009.

Jamie, Kathleen. *Sightlines: A Conversation with the Natural World*. New York: The Experiment, 2013.

Kahn, Peter H., Jr., and Patricia H. Hasbach, eds. *The Rediscovery of the Wild*. Cambridge, MA: MIT Press, 2013.

Kaye, Roger. *Last Great Wilderness: The Campaign to Establish the Arctic National Wildlife Refuge*. Fairbanks, AK: University of Alaska Press, 2006.

Keller, Robert H., and Michael F. Turek. *American Indians and National Parks*. Tucson: University of Arizona Press, 1998.

Kilgore, Bruce M., ed. *Wilderness in a Changing World*. San Francisco: Sierra Club Books, 1966.

Kimmerer, Robin Wall. *Braiding Sweetgrass: Indigenous Wisdom, Scientific Knowledge, and the Teachings of Plants*. Minneapolis, MN: Milkweed Editions, 2013.

Kolbert, Elizabeth. *Field Notes from a Catastrophe: Man, Nature, and Climate Change*. New York: Bloombsury, 2006.

————. *The Sixth Extinction: An Unnatural History*. New York: Henry Holt, 2014.

Krech III, Shepard. *The Ecological Indian: Myth and History*. New York: W. W. Norton, 1999.

Leopold, Aldo. *A Sand County Almanac: And Sketches Here and There*. Oxford, UK: Oxford University Press, 1987 edition.

Lewis, Michael, ed. *American Wilderness: A New History*. Oxford, UK: Oxford University Press, 2007.

Livingston, Dewey. *Discovering Historic Ranches at Point Reyes*. Point Reyes, CA: Point Reyes National Seashore Association, 2009.

Lopez, Barry. *Of Wolves and Men*. New York: Scribner, 1978.

———. *Arctic Dreams*. New York: Vintage, 1986.

Lyon, Ted B., and Will N. Graves. *The Real Wolf: The Science, Politics, and Economics of Co-Existing with Wolves in Modern Times*. Self-published, 2014.

MacFarlane, Robert. *Mountains of the Mind: Adventures in Reaching the Summit*. New York: Vintage, 2003.

MacKinnon, J. B. *The Once and Future World: Finding Wilderness in the Nature We've Made*. Boston: Houghton Mifflin Harcourt, 2013.

Margolin, Malcolm. *The Ohlone Way: Indian Life in the San Francisco–Monterey Bay Area*. Berkeley, CA: Heyday Books, 2003 edition.

McKibben, Bill, ed. *American Earth: Environmental Writing Since Thoreau*. New York: Literary Classics of the United States, 2008.

McMurtry, Larry. *Crazy Horse: A Life*. New York: Penguin, 1999.

McPhee, John. *Encounters with the Archdruid*. New York: Farrar, Straus and Giroux, 1971.

———. *Coming into the Country*. New York: Farrar, Straus and Giroux, 1976.

———. *Giving Good Weight*. New York: Farrar, Straus and Giroux, 1986.

———. *The Control of Nature*. New York: Farrar, Straus and Giroux, 1989.

Mann, Charles C. *1491: New Revelations of the Americas before Columbus*. New York: Vintage, 2005.

Manning, Richard. *Against the Grain: How Agriculture Has Hijacked Civilization*. New York: North Point Press / Farrar, Straus and Giroux, 2004.

Marris, Emma. *Rambunctious Garden: Saving Nature in a Post-Wild World*. New York: Bloomsbury, 2011.

Marshall III, Joseph M. *The Lakota Way: Stories and Lessons for Living*. New York: Penguin Compass, 2001.

Meyer, Stephen M. *The End of the Wild*. Sommerville, MA: Boston Review, 2006.

Monbiot, George. *Feral: Searching for Enchantment on the Frontiers of Rewilding*. London: Allen Lane, 2013.

Mooallem, Jon. *Wild Ones: A Sometimes Dismaying, Weirdly Reassuring Story about Looking at People Looking at Animals in America*. New York: Penguin, 2013.

Muir, John. *My First Summer in the Sierra*. New York: Penguin Books, 1997 edition.

———. *The Mountains of California.* Seaside, OR: Merchant Books, 2009 edition.

Nash, Roderick Frazier. *The Rights of Nature: A History of Environmental Ethics.* Madison, WI: University of Wisconsin Press, 1989.

———. *Wilderness and The American Mind.* New Haven, CT: Yale University Press, 2001 edition.

Neihardt, John G. *Black Elk Speaks.* Lincoln, NE: University of Nebraska Press, 1979 edition.

Oelschlaeger, Max. *The Idea of Wilderness.* New Haven, CT: Yale University Press, 1991.

Ostler, Jeffrey. *The Lakotas and the Black Hills: The Struggle for Sacred Ground.* New York: Penguin Books, 2010.

Pollan, Michael. *Second Nature: A Gardener's Education.* New York: Grove Press, 2003 edition.

Roberts, David. *Once They Moved Like the Wind: Cochise, Geronimo, and the Apache Wars.* New York: Touchstone, 1993.

Robinson, Michael J. *Predatory Bureaucracy: The Extermination of Wolves and the Transformation of the West.* Boulder, CO: University Press of Colorado, 2005

Ronald, Ann, ed. *Words for the Wild: The Sierra Club Trailside Reader.* San Francisco: Sierra Club Books, 1987.

Scott, Doug. *Our Wilderness: America's Common Ground.* Golden, CO: Fulcrum Publishing. 2009.

Shepard, Paul. *The Tender Carnivore and the Sacred Game.* Athens, GA: University of Georgia Press, 1973.

Snyder, Gary. *The Practice of the Wild.* San Francisco: North Point Press, 1990.

Solnit, Rebecca. *Savage Dreams: A Journey into the Landscapes of the American West.* Berkeley, CA: University of California Press, 1994.

Stegner, Wallace. *Where the Bluebird Sings to the Lemonade Springs: Living and Writing in the West.* New York: Penguin Books, 1992 edition.

———. *Marking the Sparrow's Fall: The Making of the American West,* ed. Page Stegner. New York: Henry Holt, 1998.

Stegner, Wallace, ed. *This Is Dinosaur: Echo Park Country and Its Magic Rivers.* Boulder, CO: Robert Rinehart, 1955.

Sterba, Jim. *Nature Wars: The Incredible Story of How Wildlife Comebacks Turned Backyards into Battlegrounds.* New York: Crown, 2012.

Stone, Christopher D. *Should Trees Have Standing?: Law, Morality, and the Environment*. Oxford, UK: Oxford University Press, 2010 edition.

Strayed, Cheryl. *Wild: From Lost to Found on the Pacific Crest Trail*. New York: Vintage, 2013.

Sutter, Paul S. *Driven Wild: How the Fight Against Automobiles Launched the Modern Wilderness Movement*. Seattle: University of Washington Press, 2005.

Thoreau, Henry David. *Walden and Other Writings*. New York: Modern Library, 2000 edition.

———. *The Natural History Essays*. Layton, UT: Gibbs Smith, 1988 edition.

Turner, Jack. *The Abstract Wild*. Tucson, AZ: University of Arizona Press, 1996.

Tweed, William C. *Uncertain Path: A Search for the Future of the National Parks*. Berkeley, CA: University of California Press, 2010.

White, Fred D., ed. *Essential Muir: A Selection of John Muir's Best Writings*. Berkeley, CA: Heyday, 2006.

Williams, Terry Tempest. *Refuge: An Unnatural History of Family and Place*. New York: Vintage, 1992.

Wilson, E. O. *Biophilia: The Human Bond with Other Species*. Cambridge, MA: Harvard University Press, 1984.

Wuerthner, George, Eileen Crist, and Tom Butler, eds. *Keeping the Wild: Against the Domestication of Earth*. Washington, DC: Island Press, 2014.

Zontek, Ken. *Buffalo Nation: American Indian Efforts to Restore the Bison*. Lincoln, NE: University of Nebraska Press, 2007.

Field Guides and Reference Books

As an amateur naturalist, I find it useful to have a decent field guide with me in the wilderness. These days there are many useful phone apps for the botanist and the birdwatcher. But I still prefer to go into the backcountry with a physical guide. The lightweight, laminated Pocket Naturalist Guides by Waterford Press and the Mac's Pocket Guides are both reliable. Among those in my collection that were helpful for this book, I have *Arizona Wildlife* (Waterford); *California Birds* (Waterford); *South Dakota Trees & Flowers* (Waterford); *Southwest Cacti, Trees, & Shrubs* (Mac's); *Washington State Trees & Wildflowers* (Waterford); *Wildflowers of the Southern Rocky Mountains* (Mac's).

I also have these reference guides in my library:

Angier, Bradford. *Field Guide to Edible Wild Plants*. Harrisburg, PA: Stackpole Books, 1974.

Armstrong, Robert H. *Alaska's Birds*. Anchorage, AK: Alaska Northwest Books, 2006.

Johnson, James R., and Gary E. Larson. *Grassland Plants of South Dakota and the Northern Great Plains*. Hot Springs, SD: Black Hills Parks & Forests Association, 2007.

Keator, Glenn and Ruth M. Heady. *Pacific Coast Fern Finder*, Nature Study Guide series. Birmingham, AL: Nature Study Guild, 1981.

Laws, John Muir. *The Laws Field Guide to the Sierra Nevada*. Berkeley, CA: Heyday, 2007.

Little, Elbert L. *National Audubon Society Field Guide to Trees: Western Region*. New York: Alfred A. Knopf, 1980.

Pratt, Verna E. *Alaskan Wildflowers*. Anchorage, AK: Alaskakrafts, 1990.

Sale, Richard. *A Complete Guide to Arctic Wildlife*. Buffalo, NY: Firefly, 2012.

Sibley, David Allen. *The Sibley Guide to Birds*. New York: Alfred A. Knopf, 2000.

About the Author

JASON MARK'S WRITINGS on the environment have appeared in the *New York Times*, Atlantic.com, *The Nation*, and Salon.com, among many other publications. He is the longtime editor of *Earth Island Journal*, a quarterly magazine, and a cofounder of Alemany Farm, San Francisco's largest urban farm. A long time ago, *Time* called him "a rebel with a cause."

More at jasondovemark.com

Index